170
Advances in Biochemical Engineering/Biotechnology

Aims and Scope

This book series reviews current trends in modern biotechnology and biochemical engineering. Its aim is to cover all aspects of these interdisciplinary disciplines, where knowledge, methods and expertise are required from chemistry, biochemistry, microbiology, molecular biology, chemical engineering and computer science.

Volumes are organized topically and provide a comprehensive discussion of developments in the field over the past 3–5 years. The series also discusses new discoveries and applications. Special volumes are dedicated to selected topics which focus on new biotechnological products and new processes for their synthesis and purification.

In general, volumes are edited by well-known guest editors. The series editor and publisher will, however, always be pleased to receive suggestions and supplementary information. Manuscripts are accepted in English.

In references, Advances in Biochemical Engineering/Biotechnology is abbreviated as *Adv. Biochem. Engin./Biotechnol.* and cited as a journal.

More information about this series at http://www.springer.com/series/10

Harald Seitz • Frank Stahl •
Johanna-Gabriela Walter

Editors

Catalytically Active Nucleic Acids

With contributions by

N. Alizadeh · D. Balke · J.-M. Garant · R. Hallaj ·
R. Hieronymus · C. Höbartner · F. Javadi-Zarnaghi ·
R. Jodoin · B. Juskowiak · J. Kosman · S. Müller ·
Jean-Pierre Perreault · S. Rouleau · A. Salimi · F. Stahl ·
J.-G. Walter

 Springer

Editors
Harald Seitz
Branch Bioanalytics and Bioprocesses
(IZI-BB) Biomarker Validation and
Assay Development
Fraunhofer Institute for Cell Therapy
and Immunology
Potsdam, Germany

Frank Stahl
Institut für Technische Chemie
Leibniz Universität Hannover
Hannover, Germany

Johanna-Gabriela Walter
Institut für Technische Chemie
Leibniz Universität Hannover
Hannover, Niedersachsen, Germany

ISSN 0724-6145   ISSN 1616-8542   (electronic)
Advances in Biochemical Engineering/Biotechnology
ISBN 978-3-030-29645-2   ISBN 978-3-030-29646-9   (eBook)
https://doi.org/10.1007/978-3-030-29646-9

This Springer imprint is published by the registered company Springer Nature Switzerland AG.
The registered company address is: Gewerbestrasse 11, 6330 Cham, Switzerland

Preface

We are very pleased to present the current volume of the series *Advances in Biochemical Engineering and Biotechnology* about "Catalytically Active Nucleic Acids".

The first natural enzymatic nucleic acids discovered and described almost 30 years ago were ribozymes. Their discovery led to a changed perception of nucleic acids. Nucleic acids (DNA and RNA) are not only carriers of genetic information but can also catalyze a multitude of different reactions and form a wide variety of structures, e.g. as aptamers that bind to a wide variety of target structures.

The term enzymatic nucleic acids describes nucleic acids with a catalytic activity. Ribozymes are cellular RNA molecules that mediate the cleavage and formation of phosphodiester bonds and the formation of peptide bonds. After the discovery of such natural Ribozymes numerous artificial ribozymes with altered catalytic activities were produced by in vitro and in vivo selection. Unlike ribozymes, DNAzymes do not occur in nature. Although the catalytic activity of nucleic acid enzymes is usually much slower than that of proteins, nucleic acid enzymes with comparable catalytic activity could be obtained by stringent selection processes. An advantage of nucleic acids enzymes is their small size. Furthermore, they are easier to produce and purify than proteins and can also withstand denaturation, e.g. by heat. Over the last few years, numerous reaction mechanisms have been elucidated. In addition, the number of publications with potential applications is growing. This book tries to present some aspects of nucleic acid enzymes.

We hope that the collection of chapters and the wide variety of topics will be useful not only for experts in the field but also for researchers starting to work in this area. We anticipate that this volume will provide fundamental aspects and helps to inspire students for this field.

Preparing this volume was not an easy task. Contributions from leading researchers and experts in this field have been contacted and asked for their contribution. The methodological aspects of catalytic active nucleic acids from experts with different background have been assembled. A critical discussion of the advantages and technical challenges is in the focus of this book. The chapters cover important current aspects in this field as well as recent developments.

Last but not least we would like to thank all the authors for their contributions, as well as Springer for implementation and support of this project. We would like to thank Ms Alameluh Damodharan and Prof. Thomas Scheper for their excellent support during the preparation of this volume.

Potsdam, Germany   Harald Seitz
Hannover, Germany   Frank Stahl
Hannover, Germany   Johanna-Gabriela Walter

Contents

Adv Biochem Eng Biotechnol (2020) 170: 1–20
DOI: 10.1007/10_2017_8
© Springer International Publishing AG 2017
Published online: 6 April 2017

RNA G-Quadruplexes as Key Motifs of the Transcriptome

Samuel Rouleau, Rachel Jodoin, Jean-Michel Garant, and Jean-Pierre Perreault

Abstract G-Quadruplexes are non-canonical secondary structures that can be adopted under physiological conditions by guanine-rich DNA and RNA molecules. They have been reported to occur, and to perform multiple biological functions, in the genomes and transcriptomes of many species, including humans. This chapter focuses specifically on RNA G-quadruplexes and reviews the most recent discoveries in the field, as well as addresses the upcoming challenges researchers studying these structures face.

Keywords Gene expression regulation, RNA G-quadruplex, RNA structure

Contents

S. Rouleau, R. Jodoin, J.-M. Garant, and Jean-Pierre Perreault (✉)
RNA Group/Groupe ARN, Département de Biochimie, Faculté de médecine des sciences de la santé, Pavillon de Recherche Appliquée au Cancer, Université de Sherbrooke, 3201 rue Jean-Mignault, Sherbrooke, QC, Canada, J1E 4K8
e-mail: Jean-Pierre.Perreault@USherbrooke.ca

Abbreviations

ASO	Antisense Oligonucleotides
EBV	Epstein–Barr virus
HCV	Hepatitis C virus
HIV	Human immuno-deficiency virus
HPV	Human papilloma virus
HSV	Herpes simplex virus
hTelo	Human telomeric sequence
hTERC or hTR	Human telomerase RNA component
hTERT	Human telomerase reverse transcriptase
IRES	Internal ribosome entry site
lncRNA	Long non-coding RNA
miRNA	MicroRNA
mRNA	Messenger RNA
pre-miRNA	MicroRNA precursor
RBPs	RNA binding proteins
siRNA	Small interfering RNA
TERRA	Telomeric repeat containing RNA
tiRNA	tRNA-derived, stress-induced RNA
tRNA	Transfer RNA
UTR	Untranslated region

1 Introduction

The "Big Bang" moment of the G-quadruplex field happened all the way back in 1900, when it was reported that concentrated guanylic acid solutions tend to form a gel [1]. Much later, Gellert et al. were the first to show that the guanines in these gels adopt a helical structure that is now known to be the G-quadruplex [2]. This peculiar structure, first thought to be only a laboratory curiosity, is now regarded as a crucial regulatory motif that is implicated in multiple biological functions.

Guanine-rich DNA and RNA sequences can fold into the G-quadruplex structure. The former were the first discovered and they have been more extensively studied. The latter are the subject of this chapter. RNA quadruplexes tend to be more stable than their DNA counterparts and have less topological diversity [3]. Indeed, RNA G-quadruplexes almost exclusively adopt a parallel conformation in which the four strands all have the same directionality. This observation is explained by the fact that the $2'$-hydroxyl group of the ribose locks the RNA in an *anti* conformation, which favors this parallel topology [4].

Balasubramanian's group was the first to report a biological role for an RNA G-quadruplex. They showed that a guanine-rich sequence located in the $5'$UTR of the NRAS oncogene's mRNA can fold into a G-quadruplex structure and repress the mRNA's translation [5]. Subsequently, many biological functions have been attributed to RNA G-quadruplexes [6]. Furthermore, many G-quadruplex-forming sequences have been found to be conserved over a large number of species [7, 8]. Recently, compelling evidence of the in vivo folding of RNA G-quadruplexes was provided using a structure-specific antibody [9]. This chapter aims to review both the current knowledge and the recent discoveries in the field of RNA G-quadruplexes. It also addresses the perspectives and challenges involved in the study of these structures.

2 Definition of a G-Quadruplex

The guanine base possesses two proton acceptor groups located on its Hoogsteen face, and two proton donor groups located on its Watson–Crick face (Fig. 1a) [10]. These features enable two guanines to interact in a Hoogsteen base pair through the formation of two hydrogen bonds. As the two guanines are involved in the base pair through only one of the two faces, they can still interact with another guanine to complete a planar cycle composed of four guanines linked through a total of eight hydrogen bonds [2]. This planar interaction is called a G-quartet and is usually not observable in solution because the partial negative charges of the oxygen atoms located in its center destabilize the whole structure. The formation of a G-quartet in solution requires the presence of a monovalent cation in its center to offset this partial negative charge (Fig. 1b). A cation of the appropriate radius forms an electrostatic interaction with the oxygens, increasing the stability of the structure. Among the cations found naturally in cells, potassium possesses the most appropriate radius in addition to being the most abundant intracellular cation.

Further stabilization is possible via π–π stacking of the guanines' aromatic cycles, which results in the formation of a tetrahelical structure [2]. This tetrahelix can be formed by several nucleic acids as an intermolecular G-quadruplex, or within a single nucleic acid string as an intramolecular G-quadruplex (Fig. 1c). The folding of an intramolecular G-quadruplex requires the presence of series of consecutive guanines which build the four strands of the helix. These four guanine series are spaced by three loops of variable lengths and nucleotide compositions. According to the classical definition, the loops lengths vary between 1 and 7 nucleotides (Fig. 1d).

This definition of a G-quadruplex permitted the evaluation of the number of potential G-quadruplexes present in the human genome using Quadparser, software that functions solely by pattern research [11]. This first estimation reported the presence of more than 376,000 potential G-quadruplexes, and had a large impact on the investigation of G-quadruplexes within human genomic sequences. The QGRS mapper software soon followed and provided greater flexibility in the definition of the motif. It also implemented a score which represents the sequence's likelihood of folding into a G-quadruplex [12]. This tool found the number of potential G-quadruplexes within human pre-mRNA sequences to be between 197,000 and 2,391,000 depending on the stringency of the parameters used by the algorithm [13].

Recently, a novel high throughput method of DNA G-quadruplexes detection revealed that the first estimation of the number of potential genomic G-quadruplexes was underestimated by about twofold. Precisely 716,310 distinct

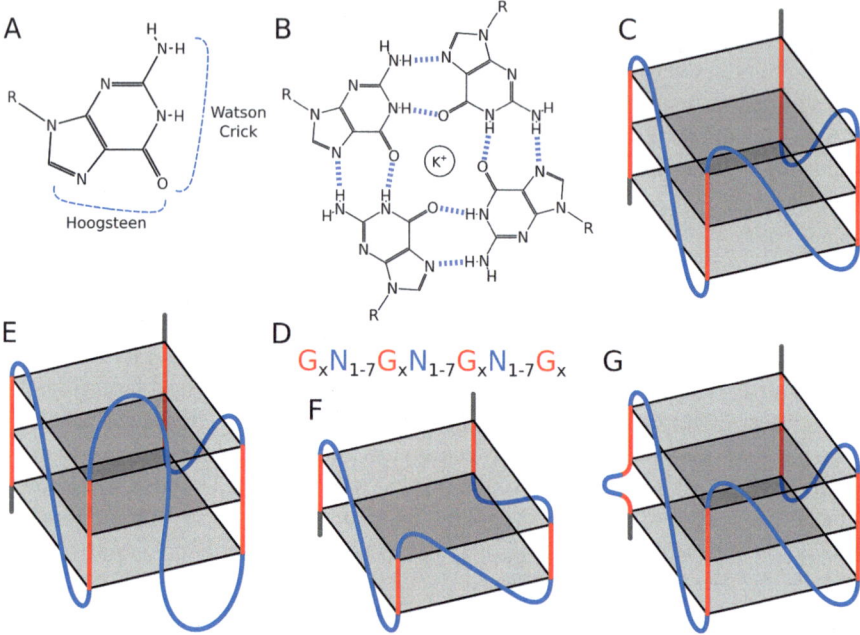

Fig. 1 The G-quadruplex structure. (**a**) Guanine base and its pairing faces. (**b**) The G-quartet. (**c**) Classical G-quadruplex. (**d**) Description of the motif originally used to find G-quadruplexes where $x \geq 3$. (**e**) G-quadruplex with a long central loop. (**f**) Two-tiered G-quadruplex. (**g**) Bulged G-quadruplex

G-Quadruplex structures were detected [14]. This discrepancy is mostly explained by G-quadruplex structures that do not respect the motif searched for by a variety of prediction software. Because such unusual structures have also been observed in RNA, the RNA G-quadruplexes estimation is thought to be an underestimation as well.

Numerous RNA G-quadruplex topologies have been reported with various combinations of guanine series and loop lengths that do not fit the previously described motif. The maximum limit of seven nucleotides per loop was shown to be incorrect, as very large second loops combined with loops one and three (which are limited to just one nucleotide each) have been observed (Fig. 1e) [15, 16]. These large loops were shown neither to be specific to the second loop nor to hinder the presence of another large loop, at least in vitro [16]. The minimum of three G-quartets implied in the motif is also erroneous. A two-quartet G-quadruplex (Fig. 1f) was shown to fold in the 5'UTR of the KRAS mRNA [17]. Additionally, G-quadruplexes that exhibit a nucleotide bulging out of a strand, which requires this nucleotide to be located within a guanine track, have been reported (Fig. 1g) [18]. An unexpected artificial G-quadruplex was observed with non-guanine bases involved in the formation of a quartet, thus widening the number of sequences that can potentially fold into G-quadruplexes [19, 20]. These multiple topologies, and the related work on DNA suggesting more structures, such as those involving triads [21], harden the challenge of identifying putative RNA G-quadruplexes.

Efforts have been made to accommodate the changing definition of what constitute a G-quadruplex in the predictions of the structure in various nucleic acids. The QGRS mapper has options to search for two quartet G-quadruplexes, as well as sequences possessing a bulge in one of the guanine series, to reduce the number of false negatives [12]. However, being more permissive increases the rate of false positive predictions and few solutions have been developed to address this problem. To provide greater confidence in its predictions, a semi-global aligner was included with the QGRS mapper in the QGRS-H predictor [22]. A potential G-quadruplex is likely to be observed if it possesses a function which should be conserved across species. The consecutive guanosine/consecutive cytosine (cG/cC) scoring system uses a different strategy by offering the user a value that reflects the presence of consecutive guanines as compared to that of consecutive cytosines. A guanine-rich environment with a low presence of cytosine is favorable to G-quadruplex formation although a balance between the two might favor Watson–Crick structures [23]. A similar strategy was used to develop G4Hunter, a program which was tested on both RNA and DNA sequences. It estimated that the number of potential G-quadruplexes in the human genome should be revised up by two- to tenfold [24]. These two tools improved G-quadruplex prediction by considering the regions flanking the potential structure.

The best strategy for identifying potential G-quadruplexes is still up for debate. The motif search provides good control of what should be considered a hit to the user. However, it is dependent on a constantly changing definition, and is therefore not appropriate for identifying new topologies of G-quadruplexes. Scoring systems favoring guanine richness correlate well with the presence of G-quadruplexes and permit the users to set their own threshold to fit their needs. However, they do not evaluate the sequence for the presence of the minimal requirements that would support intramolecular folding, and thus they promote the generation of false positives. All strategies were based on experts' observations which seem to contain a bias toward a sub-group of all possible G-quadruplexes. Fortunately, this challenge is recognized by the research community. Both the large amount of work being done and the increasing interest the field is generating should provide the volume of data (experimentally tested sequences) required to infer rules and/or patterns in the coming years. As soon as a sufficient amount of data is available, various technologies should be considered in order to learn from it. Motif detection software should obviously be used, but more advanced tools, such as computationally automated techniques, should also be considered. The latter are known to retrieve subtleties and interrelations between features that are barely noticeable by manual analysis, and are suited for use with both genomic and transcriptomic data [25].

Even though the core structure of a G-quadruplex has been thoroughly described, sequences folding into a G-quadruplex are still in need of an inclusive definition. Such a definition is almost within our grasp for intramolecular G-quadruplexes, but they represent only a fraction of all of the G-quadruplexes present in a cell. There have been reports of intermolecular G-quadruplexes involving multiple RNA molecules [26, 27]. Additionally, there have been observations of hybrid G-quadruplexes resulting from the interactions of the RNA and DNA molecules located in both telomeres and transcription forks [28, 29]. These intermolecular G-quadruplex structures force investigators to broaden the definition of what constitutes a G-quadruplex, adding another layer of complexity to the structure prediction problem. Another key challenge is to integrate G-quadruplex structure into RNA secondary structure folding software. To our knowledge, the ViennaRNA package is the only one to date that offers this option [30]. However, it relies on pattern research similar to that of Quadparser which was found to be too restrictive. It supplies a calculated evaluation of the minimal free energy of the structure based on experimental observations [31]. This evaluation allows comparison with potential Watson–Crick structures. Double-stranded structures are favored according to its predictions, but its authors are aware that the energy function of the minimal free energy was simplified in the writing of the program. It should be revised when more thermodynamics data are available, especially when considering the G-quadruplexes that do not possess three quartets, or those possessing asymmetrical loops.

3 Regulatory Roles of RNA G-Quadruplexes

RNA G-quadruplex functions have been identified at many levels of the post-transcriptional regulation of mRNA, from its transcription to its translation. Furthermore, the prediction and the identification of this structure in other families of RNA, such as miRNA, long non-coding RNA and TERRA, to name just a few, suggests that G-quadruplexes could be involved in all RNA-associated processes. This section presents an overview of the demonstrated roles of RNA G-quadruplexes, with emphasis on the most recent discoveries and on some perspectives on the global role of this regulatory motif. For more details about each G-quadruplex's functions, readers are invited to consult two recently published reviews on the subject [6, 32].

DNA G-quadruplexes located in promoters have been shown to act as transcriptional repressors [33]. Recently, Zheng et al. demonstrated not only the co-transcriptional formation of hybrid intermolecular RNA:DNA G-quadruplexes between the nascent transcript and the template DNA strand but also that these hybrid G-quadruplexes repress transcription [29]. A subsequent bioinformatic survey by the same group showed that these motifs are both conserved and highly abundant [34]. However, the best characterized roles of RNA G-quadruplexes are described in the last steps of the mRNA transcription, more specifically the termination [35] and the 3′ end-processing of the pre-mRNA [36–38]. RNA:DNA hybrids were also found to be involved in the mitochondrial transcription termination steps [39]. Overall, both RNA G-quadruplexes and RNA:DNA hybrid G-quadruplexes can be considered as *cis* regulatory elements that affect both the transcription termination and the 3′ end processing steps of polyadenylation site recognition and cleavage.

It is estimated that 92–94% of all mRNA transcripts undergo alternative splicing, creating a diversity of transcript isoforms with different roles, locations, and degrees of expression [40]. The first example of a G-quadruplex affecting alternative splicing was found in the human telomerase RNA hTERT [41]. Subsequently, RNA G-quadruplex formation and stabilization with ligands has been shown to affect the splicing of many different mRNAs, some of which are important in certain cancers [42–48]. A recent high throughput study of G-quadruplex formation in the human genome showed a high density of G-quadruplexes located near splicing sites [14]. This suggests that the splicing of many more mRNAs could be regulated by this structure. It has also been demonstrated in vitro that RNA oligonucleotides corresponding to the excised intron 1 of VEGF-A, as well as those from the introns of both PDGF A and PDGF B, could not only form G-quadruplexes but also bind to the proteins they encode. This represents yet another example of the G-quadruplex regulation of RNA splicing [49].

The specific localization of mRNA in a particular cell compartment is essential prior to its translation. G-quadruplex structures located in the 3′UTRs of both the PSD-95 and the CaMKIIa mRNAs were shown to be recognized by protein cofactors, and to be used as a signal to direct them to the neurite [50]. This aspect has been reviewed elsewhere, and lately the protein TDP-43 has been recognized as being important for the G-quadruplex mediated mRNA targeting in neurons [51, 52].

The regulation of translation is probably the most studied role of RNA G-quadruplexes. In a general way, RNA G-quadruplexes located in the 5′UTR are known to repress cap-dependent translation [15, 32, 53–57]. Conversely, there are some exceptions in which the presence of a G-quadruplex actually enhances the translation, either by being part of an IRES structure [58, 59] or by an as yet unknown mechanism [60, 61]. G-quadruplexes located in the 3′UTR have also been shown to reduce translation [62, 63]. The mechanism of the 5′UTR G-quadruplex repression of translation has not been fully determined, but two major possible mechanisms have been proposed. First, because of their high stability, they may either stall or impair the ribosome scanning of the 5′UTR. Second, they may impair the recognition of the cap structure by the eIF4e initiation factor. Further studies are also needed to decipher the exact molecular mechanism of the translational repression by G-quadruplexes located in the 3′UTR. They could recruit translation protein co-factors, affect the miRNA mediated regulation, or even influence the location of the mRNA transcript. G-quadruplexes present in the coding regions of mRNA may also impair translational elongation. Because of their high stability, they were shown to repress translation by acting as road-blocks to elongation [64, 65] and also to trigger ribosomal frame-shifting [66, 67]. The translational halt during elongation was also shown to affect the folding of the resulting protein and thus to affect its proteolytic cleavage [68]. Moreover, it has been shown that the halting effect in elongation is dependent on the G-quadruplex's exact position with respect to the open reading frame [69]. G-quadruplexes seem to be a common motif in global translational regulation. It was shown that the inhibition of eIF4A, the helicase component of the translational initiation complex, leads to the translational repression of a subset of genes, all of which are enriched in G-quadruplex motifs [70]. Interestingly, a lot of these genes were already known for their implications in various cancers. Understanding of the exact molecular mechanisms of G-quadruplex translational regulation is an essential problem to solve as more and more G-quadruplex motifs are identified and selected as potential regulatory targets. Interestingly, the regulation of translation by G-quadruplexes is well-conserved among many species, as even prokaryotes use a similar mechanism. A method has been developed to study the impact of RNA G-quadruplex motifs on mRNAs translation in *Escherichia coli* [71]. It showed that naturally occurring *E. coli* G-quadruplex motifs proximal to the ribosome binding site could affect translation [72]. Undoubtedly, RNA G-quadruplex formation could also have a tremendous impact on bacterial gene regulation and requires further investigation.

Telomere regions are highly enriched in DNA G-quadruplexes. Nevertheless, these chromosomal extremities are transcribed into the heterogeneous non-coding RNA called TERRA (telomeric repeat-containing RNA). They play a crucial role in the regulation of telomerase activity. Their sequences, multiple repeats of UUAGGG, have been shown to fold into intramolecular G-quadruplexes [73] and to form dimeric DNA/RNA hybrid G-quadruplexes [28]. Many proteins important for telomere maintenance bind to TERRA via the G-quadruplexes [74, 75]. TERRA can cause genome-wide alterations of gene expression in cancer cells [76]. Furthermore, the telomerase, the enzyme responsible for the elongation of these terminal sequences, uses an RNA template called human telomerase RNA (TERC or hTR), which was also shown to possess a G-quadruplex structure [77]. The latter was found to be unwound by the RHAU helicase (also known as DHX36) [78]. In summary, RNA G-quadruplexes play a central role in telomere homeostasis, thus making them a good potential therapeutic target as the telomerase is active in most cancers [79].

Putative G-quadruplex sequences have been found within long non-coding RNAs (lncRNA). Most of these were RNAs 200–300 nucleotides long and harbored G-quadruplex motifs possessing short loops of only 1 or 2 nucleotides, which are known to be very stable [80]. In addition, co-immunopurification and RNA binding assays revealed that RHAU helicase interacts with lncRNA BC200 [81]. By itself, BC200 does not form a G-quadruplex, but it has been shown to bind the unwound G-quadruplexes of the human telomerase RNA (hTR). These studies show that some lncRNAs are predicted to adopt G-quadruplex structures that are stable enough potentially to fold in vivo, and that different lncRNAs and helicases could regulate the folding of G-quadruplexes in other RNAs.

Both the RNA silencing mechanism and small non-coding RNAs have been shown to be crucial in the biology of the cell. MicroRNA (miRNA; 20–22 nucleotides) silence mRNA expression, and their biogenesis involves the transcription of a primary miRNA which is then cleaved in the nucleus by Drosha, creating a precursor miRNA (pre-miRNA). The pre-miRNA is subsequently exported to the cytoplasm and is further processed by the ribonuclease Dicer, at the extremities of its characteristic stem-loop, to create the mature miRNA. G-quadruplexes have been found in pre-miRNAs [82, 83]. It was shown that the G-quadruplex structure is in equilibrium with the stem-loop, and that the G-quadruplex impedes the Dicer cleavage, thus reducing the amount of mature miRNA. Moreover, it was demonstrated that the G-quadruplex present in close proximity to the miR-125a binding region in the 3′UTR of the target mRNA PSD-95 modifies its accessibility [84]. In addition, the proteins Lin28 and Nucleolin were shown to bind RNA G-quadruplexes in miRNA and mRNAs and to modulate their folding [85, 86]. It was recently shown that a mature miRNA sequence can fold into a G-quadruplex and that addition of stabilizing ligand prevents the miRNA from binding to its target [87]. G-quadruplexes were also reported in Piwi-RNA, a class of 23–30 nucleotides small non-coding RNAs involved in the silencing of retrotransposon elements in germline cells. In this case, with their protein partner, they appear to be both important landmarks and binding sites for a helicase involve in Piwi-RNA processing [88]. Finally, G-quadruplexes were also found in tiRNAs, which are produced by the cleavage of mature tRNA in the anticodon loop by the ribonuclease angiogenin [89]. The resulting 5′ fragments exert cytoprotective functions in cells by repressing translation and by triggering the formation of stress granules. In brief, G-quadruplex formation affects both the biogenesis and the functions of many kinds of small non-coding RNAs. Complementary information can be found in this review [90]. It could be interesting in the future to be able to predict their formation and to study their diverse roles in all processing intermediates, as well as in other RNA families such as small nuclear, small nucleolar, and ribosomal RNA.

G-quadruplex-forming sequences have been found in the genomes of the human immunodeficiency virus (HIV-1), the Epstein–Barr virus (EBV), papillomaviruses (HPV) [91], herpes simplex viruses (HSV) [92], and, more recently, the hepatitis C virus (HCV) [93]. In the case of HIV, G-quadruplexes have been identified at both the DNA and RNA levels. The formation of dimeric RNA G-quadruplexes from two G-rich regions of the viral genome is important in its replication [94]. In EBV, the level of EBNA1 protein expression is regulated by the presence of a G-quadruplex in its mRNA [95]. In HPV, whose proteins are coded in an overlapped fashion, the formation of RNA G-quadruplexes could affect the alternative splicing essential to their expression [91]. HSV possesses a highly-G-rich genome and was shown to fold into DNA G-quadruplexes [92]. Even if RNA G-quadruplexes have not yet been reported for this virus, unusual G-rich RNA structures have been observed, and the presence of RNA G-quadruplexes important for virus regulation is plausible [96]. Recently, Wang et al. identified a conserved G-rich sequence located in the core gene of HCV [93]. This sequence could fold into a G-quadruplex and be targeted by G-quadruplex ligands, which results in the repression of the viral replication at both the RNA and the protein levels. Overall, G-quadruplex formation in viruses seems to be very important for both their replication and their gene expression, and is considered a promising therapeutic target going forward [97].

In light of these findings, it seems that most if not all aspects of RNA biology can be regulated by G-quadruplexes. As new classes of RNA, or new roles for known

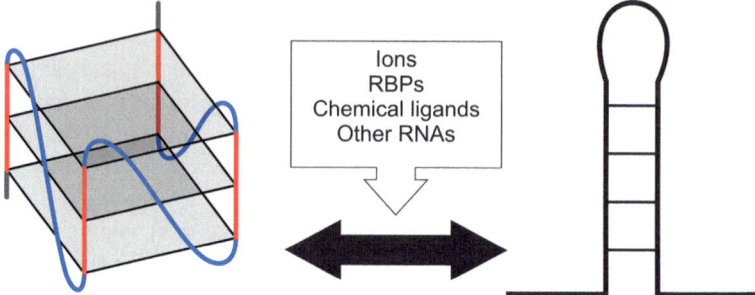

Fig. 2 Modulation of G-quadruplex folding. Different modulators used by either the cell or researchers in order to influence the folding balance between G-quadruplex and canonical structures

RNAs, are discovered, additional layers of G-quadruplex mediated regulation are sure to follow.

4 Modulation of G-Quadruplex Folding and Its Biotechnological Applications

RNA folding is a dynamic process. One given sequence can fold into several distinct structures. Therefore, in a cell, the multiple copies of an RNA molecule can reach equilibrium between alternating structures. Given that these changes can affect the RNA's function, the balance between the alternate structures can be used both to modulate and to fine-tune the function of a given RNA (Fig. 2). Many factors can change RNA folding, including temperature, ionic concentrations, and the binding of a metabolite, a protein, or another RNA molecule, thus adding another layer to RNA regulation (reviewed in [98]).

Competition between canonical Watson–Crick and G-quadruplex folding is a well-documented phenomenon [23, 53, 82, 83, 99]. It has been shown that the presence of consecutive cytosines in the vicinity of the G-runs prevents quadruplex folding [23, 53]. Conversely, in the past year, two groups have independently shown that the presence of G-runs in pre-miRNAs leads to quadruplex folding and prevents processing by Dicer [82, 83]. The competition between a hairpin and a G-quadruplex structure was studied at various Mg^{2+} (which favors the hairpin) and K^+ (which favors the G-quadruplex) concentrations, and showed that, under physiological conditions, the quadruplex was the main form present [99]. However, it should be noted that only one sequence was tested, and that dissimilar hairpins or G-quadruplexes might have different relative stabilities and thus behave oppositely. Furthermore, in some precise situations, such as neuronal polarization/depolarization, K^+ concentrations can vary greatly. As G-quadruplexes are known to be important for the specific localization of some mRNAs to the neurite, some have speculated that the G-quadruplexes are used as a K^+ concentration sensor and thus regulate both mRNA location and translation according to neural activity [50]. This interesting hypothesis remains to be tested. Another interesting case occurs in plant cells, where K^+ deficiency leads to major changes in the expression levels of multiple genes [100]. As several plant genomes contain putative G-quadruplex sequences [101–104], it is tempting to assume that these G-quadruplexes could play a role in the K^+ homeostasis in plants, but further studies are required to confirm this.

Another manner in which the cell can regulate RNA folding is to use RNA-binding proteins (RBPs). Many RBPs have been shown to bind to G-quadruplex structures [49, 57, 65, 105]. Some were shown to unwind RNA G-quadruplexes, including RHAU [106], lin28 [85], DHX9 [107], and eIF4A [108]. Among these, RHAU is the best characterized [78, 109–111]. Many other RBPs/G-quadruplex interactions are notably less well-studied. Furthermore, given the high number of both RBPs and G-quadruplex-forming sequences, it is highly probable that much of the RBPs/G-quadruplexes interactome remains unknown. With the advent of high throughput screening technologies, more interactions are still to be discovered. Co-immunoprecipitation, followed by RNA deep sequencing, can lead to the discovery of new RNA targets for known G-quadruplex binding proteins [81, 112]. On the other hand, G-quadruplex-forming RNAs can be used to pull down proteins which can then be identified by mass spectrometry [49, 55]. By knowing its binding partners, researchers can gain new insights into either an RBP's or an RNA G-quadruplex's functions. It is then conceivable to affect G-quadruplex-related functions by targeting the desired RBPs. In this way, a subset of G-quadruplexes involved in a particular pathway could be targeted via the right RBP.

Many chemical ligands possessing a high affinity for G-quadruplex structure have been developed. They also possess high selectivity toward the quadruplex over other nucleic acid structures [113]. As G-quadruplexes are enriched in proto-oncogenes, some of these ligands are considered to be good therapeutic candidates with which to treat various cancers [114]. However, the selectivity of a ligand for a given subclass of G-quadruplexes, including the selectivity for RNA over DNA and even for a single sequence, is the greatest challenge in the field of G-quadruplex ligands as off-target effects remain an issue [113]. Another problem is the fact that these ligands could induce G-quadruplex folding in sequences that would not adopt this structure under normal conditions [14, 115].

Alternatively, engineered proteins can be used to bind and study G-quadruplexes. A G-quadruplex-specific antibody has been used to visualize these RNA structures in live cells [9]. In another study, the RNA binding domain of the TLS protein was used to create two distinct proteins that were able both to discriminate and to specifically bind to the hTelo and the TERRA G-quadruplexes, the DNA and RNA telomeric sequences, respectively. This enabled the authors to study the distinct effects that these two G-quadruplexes have on telomeric DNA chromatin's status [116]. As more quadruplex binding proteins are found, and the development of protein engineering continues [117], more of these valuable tools should become available for the study and targeting of RNA G-quadruplexes.

Antisense oligonucleotides (ASO) are yet another tool that can be used to target G-quadruplexes. Their main advantage is their specificity, as they recognize their target via Watson–Crick base pairing. Moreover, they have a great plasticity, as many chemical modifications of the ASO backbone and/or sugar moieties can be used to alter the pharmacokinetics properties, stability, specificity, and affinity of the ASO, so as to suit specific needs [118]. Besides, ASO can be used to modulate G-quadruplex folding in both directions. ASO can target guanine stretches and prevent quadruplex folding, thus leading to enhanced translation of specific mRNAs [95, 119]. Alternatively, ASO can target G-quadruplex neighboring sequences, thus interfering with competing canonic structures and enhancing quadruplex folding [119]. Other ASO can invade existing quadruplexes and form hetero G-quadruplex structures, leading to a decrease in mRNA translation [120]. Some can even form a hetero quadruplex with sequences that do not fold into G-quadruplexes on their own, leading to either a decrease in translation [121] or a hindrance of viral replication [122]. Differences in both the length and the chemical composition of the ASO, as well as the targeted sequence, can lead to different outcomes [123]. Further optimization of these parameters should lead to the generation of efficient ASO possessing a wide range of effects. The major hurdle preventing the use of ASO as therapeutic agents is their delivery to specific cells. Fortunately, a lot of research has been done on this problem for siRNAs [124–127], and the lessons learned there can be useful for the delivery of ASO.

It is also possible to use more than one quadruplex binder at a time [128]. Future research should focus on the creation of chimeric ASO/chemical ligands/protein compounds that can combine the different quadruplex binders' respective strengths, compensating for their respective weaknesses. Bifunctional oligonucleotides composed of an antisense portion that determines target specificity, and a non-hybridizing tail that recruits proteins or RNA/protein complexes, have been successfully used to modulate specific mRNA splicing [129]. A similar strategy could be used with a quadruplex folding tail to recruit quadruplex binding proteins to different targets and thus regulate different biological processes. ASO could also be fused to quadruplex ligands in order to enhance their specificity. Given the high prevalence of quadruplex-forming sequences, and their implication in a plethora of biological functions, the imagination and creativity of researchers seem to be the only limits to the potential usefulness of G-quadruplex modulation.

Clearly, there are many ways to manipulate the equilibrium between G-quadruplexes and other secondary structures. One great technical challenge that remains is to find a way to quantify the proportion of a given RNA that folds into each of its various structures in solution, or even within a cell.

Additionally, RNA G-quadruplexes are important in other biotechnological applications. For example, two different RNA aptamers designed for binding and activating ligand fluorescence, namely RNA mango [130] and RNA spinach [19, 20], were both shown to bind to their respective ligand with a G-quadruplex structure located within their binding pockets. Interestingly, many known DNA and RNA aptamers fold into G-quadruplex structures [131, 132]. This exciting field is sure to develop in the near future as more knowledge on quadruplexes becomes available.

5 Conclusion

The RNA G-quadruplex field has been expanding quickly in the last few years, but these fascinating structures have yet to reveal all their secrets. As more and more tools become available for their study, more functions can be identified. In parallel, the number of possible applications and biotechnological tools utilizing them continue to grow. RNA G-quadruplexes are sure to keep scientists from many fields busy for years to come.

References

1. Bang I (1900) Chemische und physiologische Studien über die guanylsäure. I. Theil. chemische studien. Hoppe Seylers Z Physiol Chem 31:411–427
2. Gellert M, Lipsett MN, Davies DR (1962) Helix formation by guanylic acid. Proc Natl Acad Sci U S A 48:2013–2018
3. Joachimi A, Benz A, Hartig JS (2009) A comparison of DNA and RNA quadruplex structures and stabilities. Bioorg Med Chem 17(19):6811–6815
4. Tang C-F, Shafer RH (2006) Engineering the quadruplex fold: nucleoside conformation determines both folding topology and molecularity in guanine quadruplexes. J Am Chem Soc 128: 5966–5973
5. Kumari S, Bugaut A, Huppert JL, Balasubramanian S (2007) An RNA G-quadruplex in the 5'UTR of the NRAS proto-oncogene modulates translation. Nat Chem Biol 3(4):218–221
6. Millevoi S, Moine H, Vagner S (2012) G-quadruplexes in RNA biology. Wiley Interdiscip Rev 3(4):495–507
7. Yadav VK, Abraham JK, Mani P, Kulshrestha R, Chowdhury S (2008) QuadBase: genome-wide database of G4 DNA–occurrence and conservation in human, chimpanzee, mouse and rat promoters and 146 microbes. Nucleic Acids Res 36(Database issue):D381–D385
8. Frees S, Menendez C, Crum M, Bagga PS (2014) QGRS-Conserve: a computational method for discovering evolutionarily conserved G-quadruplex motifs. Hum Genomics 8:8
9. Biffi G, Di Antonio M, Tannahill D, Balasubramanian S (2014) Visualization and selective chemical targeting of RNA G-quadruplex structures in the cytoplasm of human cells. Nat Chem 6(1):75–80
10. Hoogsteen K (1963) The crystal and molecular structure of a hydrogen-bonded complex between 1-methylthymine and 9-methyladenine. Acta Crystallogr 16(9):907–916
11. Huppert JL, Balasubramanian S (2005) Prevalence of quadruplexes in the human genome. Nucleic Acids Res 33(9):2908–2916
12. Kikin O, D'Antonio L, Bagga PS (2006) QGRS mapper: a web-based server for predicting G-quadruplexes in nucleotide sequences. Nucleic Acids Res 34(Web Server issue): W676–W682
13. Kikin O, Zappala Z, D'Antonio L, Bagga PS (2007) GRSDB2 and GRS_UTRdb: databases of quadruplex forming G-rich sequences in pre-mRNAs and mRNAs: databases of quadruplex forming G-rich sequences in pre-mRNAs and mRNAs. Nucleic Acids Res 36 (Database):D141–D148

14. Chambers VS, Marsico G, Boutell JM, Di Antonio M, Smith GP, Balasubramanian S (2015) High-throughput sequencing of DNA G-quadruplex structures in the human genome. Nat Biotechnol 33(8):877–881

15. Jodoin R, Bauer L, Garant J-M, Mahdi Laaref A, Phaneuf F, Perreault J-P (2014) The folding of 5′-UTR human G-quadruplexes possessing a long central loop. RNA 20(7):1129–1141

16. Pandey S, Agarwala P, Maiti S (2013) Effect of loops, G-quartets on the stability of RNA G-quadruplexes. J Phys Chem 117:6896–6905

17. Faudale M, Cogoi S, Xodo LE (2011) Photoactivated cationic alkyl-substituted porphyrin binding to g4-RNA in the 5′-UTR of KRAS oncogene represses translation. Chem Commun (Camb) 48:874–876

18. Martadinata H, Phan AT (2014) Formation of a stacked dimeric G-quadruplex containing bulges by the 5′-terminal region of human telomerase RNA (hTERC). Biochemistry 53:1595–1600

19. Huang H, Suslov NB, Li N-S, Shelke SA, Evans ME, Koldobskaya Y, Rice PA, Piccirilli JA (2014) A G-quadruplex-containing RNA activates fluorescence in a GFP-like fluorophore. Nat Chem Biol 10:686–691

20. Warner KD, Chen MC, Song W, Strack RL, Thorn A, Jaffrey SR, Ferré-D'Amaré AR (2014) Structural basis for activity of highly efficient RNA mimics of green fluorescent protein. Nat Struct Mol Biol 21(8):658–663

21. Heddi B, Martín-Pintado N, Serimbetov Z, Kari TMA, Phan AT (2016) G-quadruplexes with (4n − 1) guanines in the G-tetrad core: formation of a G-triad·water complex and implication for small-molecule binding. Nucleic Acids Res 44(2):910–916

22. Menendez C, Frees S, Bagga PS (2012) QGRS-H Predictor: a web server for predicting homologous quadruplex forming G-rich sequence motifs in nucleotide sequences. Nucleic Acids Res 40(Web Server issue):W96–W103

23. Beaudoin J-D, Jodoin R, Perreault J-P (2014) New scoring system to identify RNA G-quadruplex folding. Nucleic Acids Res 42(2):1209–1223

24. Bedrat A, Lacroix L, Mergny J-L (2016) Re-evaluation of G-quadruplex propensity with G4Hunter. Nucleic Acids Res 44(4):1746–1759

25. Libbrecht MW, Noble WS (2015) Machine learning applications in genetics and genomics. Nat Rev Genet 16(6):321–332

26. Martadinata H, Phan AT (2009) Structure of propeller-type parallel-stranded RNA G-quadruplexes, formed by human telomeric RNA sequences in K^+ solution. J Am Chem Soc 131(7):2570–2578

27. Collie GW, Parkinson GN, Neidle S, Rosu F, De Pauw E, Gabelica V (2010) Electrospray mass spectrometry of telomeric RNA (TERRA) reveals the formation of stable multimeric G-quadruplex structures. J Am Chem Soc 132(27):9328–9334

28. Xu Y, Kimura T, Komiyama M (2007) Human telomere RNA and DNA form an intermolecular G-quadruplex. Nucleic Acids Symp Ser 52:169–170

29. Zheng K, Xiao S, Liu J, Zhang J, Hao Y, Tan Z (2013) Co-transcriptional formation of DNA: RNA hybrid G-quadruplex and potential function as constitutional cis element for transcription control. Nucleic Acids Res 41(10):5533–5541

30. Lorenz R, Bernhart SH, Qin J, Höner zu C, Siederdissen AT, Amman F, Hofacker IL, Stadler PF (2013) 2D meets 4G: G-quadruplexes in RNA secondary structure prediction. IEEE/ACM Trans Comput Biol Bioinforma/IEEE, ACM 10(4):832–844

31. Zhang AYQ, Bugaut A, Balasubramanian S (2011) A sequence-independent analysis of the loop length dependence of intramolecular RNA G-quadruplex stability and topology. Biochemistry 50(33):7251–7258

32. Agarwala P, Pandey S, Maiti S (2015) The tale of RNA G-quadruplex. Org Biomol Chem 13: 5570–5585

33. Huppert LJ, Balasubramanian S (2007) G-quadruplexes in promoters throughout the human genome. Nucleic Acids Res 35:406–413

34. Xiao S, Zhang J-Y, Zheng K-W, Hao Y-H, Tan Z (2013) Bioinformatic analysis reveals an evolutional selection for DNA: RNA hybrid G-quadruplex structures as putative transcription regulatory elements in warm-blooded animals. Nucleic Acids Res 41:10379–10390
35. Wanrooij HP, Uhler PJ, Simonsson T, Falkenberg M, Gustafsson MC (2010) G-quadruplex structures in RNA stimulate mitochondrial transcription termination and primer formation. Proc Natl Acad Sci U S A 107:16072–16077
36. Christiansen J, Kofod M, Nielsen CF (1994) A guanosine quadruplex and two stable hairpins flank a major cleavage site in insulin-like growth factor II mRNA. Nucleic Acids Res 22: 5709–5716
37. Decorsiere A, Cayrel A, Vagner S, Millevoi S (2011) Essential role for the interaction between hnRNP H/F and a G quadruplex in maintaining p53 pre-mRNA 3′-end processing and function during DNA damage. Genes Dev 25(3):220–225
38. Beaudoin J-D, Perreault J-P (2013) Exploring mRNA 3′-UTR G-quadruplexes: evidence of roles in both alternative polyadenylation and mRNA shortening. Nucleic Acids Res 41(11): 5898–5911
39. Zheng K, Wu R, He Y, Xiao S, Zhang J, Liu J, Hao Y, Tan Z (2014) A competitive formation of DNA: RNA hybrid G-quadruplex is responsible to the mitochondrial transcription termination at the DNA replication priming site. Nucleic Acids Res 42:10832–10844
40. Wang TE, Sandberg R, Luo S, Khrebtukova I, Zhang L, Mayr C, Kingsmore FS, Schroth PG, Burge BC (2008) Alternative isoform regulation in human tissue transcriptomes. Nature 456: 470–476
41. Gomez D, Lemarteleur T, Lacroix L, Mailliet P, Mergny J-L, Riou J-F (2004) Telomerase downregulation induced by the G-quadruplex ligand 12459 in A549 cells is mediated by hTERT RNA alternative splicing. Nucleic Acids Res 32:371–379
42. Fisette J-F, Montagna DR, Mihailescu M-R, Wolfe MS (2012) A G-rich element forms a G-quadruplex and regulates BACE1 mRNA alternative splicing. J Neurochem 121(5): 763–773
43. Ribeiro MM, Teixeira SG, Martins L, Marques RM, de Souza PA, Line PSR (2015) G-quadruplex formation enhances splicing efficiency of PAX9 intron 1. Hum Genet 134: 37–44
44. Hai Y, Cao W, Liu G, Hong S-P, Elela AS, Klinck R, Chu J, Xie J (2008) A G-tract element in apoptotic agents-induced alternative splicing. Nucleic Acids Res 36:3320–3331
45. Zizza P, Cingolani C, Artuso S, Salvati E, Rizzo A, D'Angelo C, Porru M, Pagano B, Amato J, Randazzo A, Novellino E, Stoppacciaro A, Gilson E, Stassi G, Leonetti C, Biroccio A (2016) Intragenic G-quadruplex structure formed in the human CD133 and its biological and translational relevance. Nucleic Acids Res 44:1579–1590
46. Marcel V, Tran PL, Sagne C, Martel-Planche G, Vaslin L, Teulade-Fichou MP, Hall J, Mergny JL, Hainaut P, Van Dyck E (2011) G-quadruplex structures in TP53 intron 3: role in alternative splicing and in production of p53 mRNA isoforms. Carcinogenesis 32(3):271–278
47. Perriaud L, Marcel V, Sagne C, Favaudon V, Guédin A, De Rache A, Guetta C, Hamon F, Teulade-Fichou M-P, Hainaut P, Mergny J-L, Hall J (2014) Impact of G-quadruplex structures and intronic polymorphisms rs17878362 and rs1642785 on basal and ionizing radiation-induced expression of alternative p53 transcripts. Carcinogenesis 35:2706–2715
48. Didiot M-C, Tian Z, Schaeffer C, Subramanian M, Mandel J-L, Moine H (2008) The G-quartet containing FMRP binding site in FMR1 mRNA is a potent exonic splicing enhancer. Nucleic Acids Res 36:4902–4912
49. Saito T, Yoshida W, Yokoyama T, Abe K, Ikebukuro K (2015) Identification of RNA oligonucleotides binding to several proteins from potential G-quadruplex forming regions in transcribed pre-mRNA. Molecules 20(11):20832–20840
50. Subramanian M, Rage F, Tabet R, Flatter E, Mandel JL, Moine H (2011) G-quadruplex RNA structure as a signal for neurite mRNA targeting. EMBO Rep 12(7):697–704

51. Schofield RJP, Cowan LJ, Coldwell JM (2015) G-quadruplexes mediate local translation in neurons. Biochem Soc Trans 43:338–342
52. Ishiguro A, Kimura N, Watanabe Y, Watanabe S, Ishihama A (2016) TDP-43 binds and transports G-quadruplex-containing mRNAs into neurites for local translation. Genes Cells 21:466–481
53. Beaudoin JD, Perreault JP (2010) 5′-UTR G-quadruplex structures acting as translational repressors. Nucleic Acids Res 38(20):7022–7036
54. Lammich S, Kamp F, Wagner J, Nuscher B, Zilow S, Ludwig A-K, Willem M, Haass C (2011) Translational repression of the disintegrin and metalloprotease ADAM10 by a stable G-quadruplex secondary structure in its 5′-untranslated region. J Biol Chem 286: 45063–45072
55. von Hacht A, Seifert O, Menger M, Schütze T, Arora A, Konthur Z, Neubauer P, Wagner A, Weise C, Kurreck J (2014) Identification and characterization of RNA guanine-quadruplex binding proteins. Nucleic Acids Res 42:6630–6644
56. Nie J, Jiang M, Zhang X, Tang H, Jin H, Huang X, Yuan B, Zhang C, Lai CJ, Nagamine Y, Pan D, Wang W, Yang Z (2015) Post-transcriptional Regulation of Nkx2-5 by RHAU in heart development. Cell Rep 13:723–732
57. Williams KR, McAninch DS, Stefanovic S, Xing L, Allen M, Li W, Feng Y, Mihailescu MR, Bassell GJ (2016) hnRNP-Q1 represses nascent axon growth in cortical neurons by inhibiting Gap-43 mRNA translation. Mol Biol Cell 27(3):518–534
58. Bonnal S, Schaeffer C, Créancier L, Clamens S, Moine H, Prats A-C, Vagner S (2003) A single internal ribosome entry site containing a G quartet RNA structure drives fibroblast growth factor 2 gene expression at four alternative translation initiation codons. J Biol Chem 278:39330–39336
59. Morris MJ, Negishi Y, Pazsint C, Schonhoft JD, Basu S (2010) An RNA G-quadruplex is essential for cap-independent translation initiation in human VEGF IRES. J Am Chem Soc 132(50):17831–17839
60. Agarwala P, Pandey S, Mapa K, Maiti S (2013) The G-quadruplex augments translation in the 5′ untranslated region of transforming growth factor β2. Biochemistry 52:1528–1538
61. Agarwala P, Pandey S, Maiti S (2014) Role of G-quadruplex located at 5′ end of mRNAs. Biochim Biophys Acta 1840:3503–3510
62. Arora A, Suess B (2011) An RNA G-quadruplex in the 3′ UTR of the proto-oncogene PIM1 represses translation. RNA Biol 8:802–805
63. Crenshaw E, Leung PB, Kwok KC, Sharoni M, Olson K, Sebastian PN, Ansaloni S, Schweitzer-Stenner R, Akins RM, Bevilacqua CP, Saunders JA (2015) Amyloid precursor protein translation is regulated by a 3′UTR guanine quadruplex. PLoS One 10:e0143160
64. Endoh T, Kawasaki Y, Sugimoto N (2013) Suppression of gene expression by G-quadruplexes in open reading frames depends on G-quadruplex stability. Angew Chem 52:5522–5526
65. Thandapani P, Song J, Gandin V, Cai Y, Rouleau SG, Garant J-M, Boisvert F-M, Yu Z, Perreault J-P, Topisirovic I, Richard S (2015) Aven recognition of RNA G-quadruplexes regulates translation of the mixed lineage leukemia protooncogenes. elife 4:e06234
66. Endoh T, Sugimoto N (2013) Unusual -1 ribosomal frameshift caused by stable RNA G-quadruplex in open reading frame. Anal Chem 85:11435–11439
67. Yu C-H, Teulade-Fichou M-P, Olsthoorn LRC (2014) Stimulation of ribosomal frameshifting by RNA G-quadruplex structures. Nucleic Acids Res 42:1887–1892
68. Endoh T, Kawasaki Y, Sugimoto N (2013) Stability of RNA quadruplex in open reading frame determines proteolysis of human estrogen receptor? Nucleic Acids Res 41:6222–6231
69. Endoh T, Sugimoto N (2016) Mechanical insights into ribosomal progression overcoming RNA G-quadruplex from periodical translation suppression in cells. Sci Report 6:22719
70. Raza F, Waldron AJ, Quesne LJ (2015) Translational dysregulation in cancer: eIF4A isoforms and sequence determinants of eIF4A dependence. Biochem Soc Trans 43: 1227–1233

71. Wieland M, Hartig SJ (2009) Investigation of mRNA quadruplex formation in *Escherichia coli*. Nat Protoc 4:1632–1640
72. Holder TI, Hartig SJ (2014) A matter of location: influence of G-quadruplexes on *Escherichia coli* gene expression. Chem Biol 21:1511–1521
73. Martadinata H, Heddi B, Lim WK, Phan TA (2011) Structure of long human telomeric RNA (TERRA): G-quadruplexes formed by four and eight UUAGGG repeats are stable building blocks. Biochemistry 50:6455–6461
74. Biffi G, Tannahill D, Balasubramanian S (2012) An intramolecular G-quadruplex structure is required for binding of telomeric repeat-containing RNA to the telomeric protein TRF2. J Am Chem Soc 134:11974–11976
75. Takahama K, Oyoshi T (2013) Specific binding of modified RGG domain in TLS/FUS to G-quadruplex RNA: tyrosines in RGG domain recognize 2′-OH of the riboses of loops in G-quadruplex. J Am Chem Soc 135:18016–18019
76. Hirashima K, Seimiya H (2015) Telomeric repeat-containing RNA/G-quadruplex-forming sequences cause genome-wide alteration of gene expression in human cancer cells in vivo. Nucleic Acids Res 43:2022–2032
77. Gros J, Guédin A, Mergny J-L, Lacroix L (2008) G-Quadruplex formation interferes with P1 helix formation in the RNA component of telomerase hTERC. ChemBioChem 9:2075–2079
78. Booy EP, McRae EKS, McKenna SA (2014) Biochemical characterization of G4 quadruplex telomerase RNA unwinding by the RNA helicase RHAU. Methods Mol Biol 1259:125–135
79. Neidle S (2016) Quadruplex nucleic acids as novel therapeutic targets. J Med Chem 59:5987
80. Jayaraj GG, Pandey S, Scaria V, Maiti S (2012) Potential G-quadruplexes in the human long non-coding transcriptome. RNA Biol 9:81–89
81. Booy EP, McRae EKS, Howard R, Deo SR, Ariyo EO, Dzananovic E, Meier M, Stetefeld J, McKenna SA (2016) RNA helicase associated with AU-rich element (RHAU/DHX36) interacts with the 3′-tail of the long non-coding RNA BC200 (BCYRN1). J Biol Chem 291(10):5355–5372
82. Mirihana Arachchilage G, Dassanayake CA, Basu S (2015) A potassium ion-dependent RNA structural switch regulates human pre-miRNA 92b maturation. Chem Biol 22:262–272
83. Pandey S, Agarwala P, Jayaraj GG, Gargallo R, Maiti S (2015) The RNA stem-loop to G-quadruplex equilibrium controls mature MicroRNA production inside the cell. Biochemistry 54(48):7067–7078
84. Stefanovic S, Bassell GJ, Mihailescu MR (2015) G quadruplex RNA structures in PSD-95 mRNA: potential regulators of miR-125a seed binding site accessibility. RNA 21(1):48–60
85. O'Day E, Le MTN, Imai S, Tan SM, Kirchner R, Arthanari H, Hofmann O, Wagner G, Lieberman J (2015) An RNA-binding protein, lin28, recognizes and remodels G-quartets in the microRNAs (miRNAs) and mRNAs it regulates. J Biol Chem 290(29):17909–17922
86. Woo H-H, Baker T, Laszlo C, Chambers SK (2013) Nucleolin mediates microRNA-directed CSF-1 mRNA deadenylation but increases translation of CSF-1 mRNA. Mol Cell Proteomics 12(6):1661–1677
87. Tan W, Zhou J, Gu J, Xu M, Xu X, Yuan G (2016) Probing the G-quadruplex from hsa-miR-3620-5p and inhibition of its interaction with the target sequence. Talanta 154:560–566
88. Vourekas A, Zheng K, Fu Q, Maragkakis M, Alexiou P, Ma J, Pillai SR, Mourelatos Z, Wang JP (2015) The RNA helicase MOV10L1 binds piRNA precursors to initiate piRNA processing. Genes Dev 29:617–629
89. Ivanov P, O'Day E, Emara MM, Wagner G, Lieberman J, Anderson P (2014) G-quadruplex structures contribute to the neuroprotective effects of angiogenin-induced tRNA fragments. Proc Natl Acad Sci U S A 111:18201–18206
90. Simone R, Fratta P, Neidle S, Parkinson NG, Isaacs MA (2015) G-quadruplexes: emerging roles in neurodegenerative diseases and the non-coding transcriptome. FEBS Lett 589: 1653–1668
91. Métifiot M, Amrane S, Litvak S, Andreola M-L (2014) G-quadruplexes in viruses: function and potential therapeutic applications. Nucleic Acids Res 42:12352–12366

92. Artusi S, Nadai M, Perrone R, Biasolo AM, Palù G, Flamand L, Calistri A, Richter NS (2015) The Herpes simplex virus-1 genome contains multiple clusters of repeated G-quadruplex: Implications for the antiviral activity of a G-quadruplex ligand. Antivir Res 118:123–131

93. Wang S-R, Min Y-Q, Wang J-Q, Liu C-X, Fu B-S, Wu F, Wu L-Y, Qiao Z-X, Song Y-Y, Xu G-H, Wu Z-G, Huang G, Peng N-F, Huang R, Mao W-X, Peng S, Chen Y-Q, Zhu Y, Tian T, Zhang X-L, Zhou X (2016) A highly conserved G-rich consensus sequence in hepatitis C virus core gene represents a new anti-hepatitis C target. Sci Adv 2:e1501535–e1501535

94. Piekna-Przybylska D, Sullivan AM, Sharma G, Bambara AR (2014) U3 region in the HIV-1 genome adopts a G-quadruplex structure in its RNA and DNA sequence. Biochemistry 53: 2581–2593

95. Murat P, Zhong J, Lekieffre L, Cowieson NP, Clancy JL, Preiss T, Balasubramanian S, Khanna R, Tellam J (2014) G-quadruplexes regulate Epstein-Barr virus-encoded nuclear antigen 1 mRNA translation. Nat Chem Biol 10:358–364

96. Horsburgh CB, Kollmus H, Hauser H, Coen MD (1996) Translational recoding induced by G-rich mRNA sequences that form unusual structures. Cell 86:949–959

97. Harris ML, Merrick JC (2015) G-quadruplexes in pathogens: a common route to virulence control? PLoS Pathog 11:e1004562

98. Wan Y, Kertesz M, Spitale RC, Segal E, Chang HY (2011) Understanding the transcriptome through RNA structure. Nat Rev Genet 12(9):641–655

99. Bugaut A, Murat P, Balasubramanian S (2012) An RNA hairpin to G-quadruplex conformational transition. J Am Chem Soc 134:19953–19956

100. Amtmann A, Rubio F (2012) Potassium in plants. In: eLS. Wiley, Chichester

101. Mullen MA, Olson KJ, Dallaire P, Major F, Assmann SM, Bevilacqua PC (2010) RNA G-Quadruplexes in the model plant species Arabidopsis thaliana: prevalence and possible functional roles. Nucleic Acids Res 38(22):8149–8163

102. Andorf CM, Kopylov M, Dobbs D, Koch KE, Stroupe ME, Lawrence CJ, Bass HW (2014) G-quadruplex (G4) motifs in the maize (Zea mays L.) genome are enriched at specific locations in thousands of genes coupled to energy status, hypoxia, low sugar, and nutrient deprivation. J Genet Genomics 41(12):627–647

103. Yang M, Wu Y, Jin S, Hou J, Mao Y, Liu W, Shen Y, Wu L (2015) Flower bud transcriptome analysis of Sapium sebiferum (Linn.) Roxb. and primary investigation of drought induced flowering: pathway construction and G-quadruplex prediction based on transcriptome. PLoS One 10:e0118479

104. Wang Y, Zhao M, Zhang Q, Zhu G-F, Li F-F, Du L-F (2015) Genomic distribution and possible functional roles of putative G-quadruplex motifs in two subspecies of Oryza sativa. Comput Biol Chem 56:122–130

105. Brázda V, Hároníková L, Liao JCC, Fojta M (2014) DNA and RNA quadruplex-binding proteins. Int J Mol Sci 15(10):17493–17517

106. Creacy SD, Routh ED, Iwamoto F, Nagamine Y, Akman SA, Vaughn JP (2008) G4 resolvase 1 binds both DNA and RNA tetramolecular quadruplex with high affinity and is the major source of tetramolecular quadruplex G4-DNA and G4-RNA resolving activity in HeLa cell lysates. J Biol Chem 283:34626–34634

107. Chakraborty P, Grosse F (2011) Human DHX9 helicase preferentially unwinds RNA-containing displacement loops (R-loops) and G-quadruplexes. DNA Repair 10(6): 654–665

108. Wolfe AL, Singh K, Zhong Y, Drewe P, Rajasekhar VK, Sanghvi VR, Mavrakis KJ, Jiang M, Roderick JE, Van der Meulen J, Schatz JH, Rodrigo CM, Zhao C, Rondou P, de Stanchina E, Teruya-Feldstein J, Kelliher MA, Speleman F, Porco JA, Pelletier J, Rätsch G, Wendel H-G (2014) RNA G-quadruplexes cause eIF4A-dependent oncogene translation in cancer. Nature 513:65–70

109. Heddi B, Cheong VV, Martadinata H, Phan AT (2015) Insights into G-quadruplex specific recognition by the DEAH-box helicase RHAU: solution structure of a peptide-quadruplex complex. Proc Natl Acad Sci U S A 112:9608–9613

110. Chen MC, Murat P, Abecassis K, Ferré-D'Amaré AR, Balasubramanian S (2015) Insights into the mechanism of a G-quadruplex-unwinding DEAH-box helicase. Nucleic Acids Res 43: 2223–2231

111. Ariyo EO, Booy EP, Patel TR, Dzananovic E, McRae EK, Meier M, McEleney K, Stetefeld J, McKenna SA (2015) Biophysical characterization of G-quadruplex recognition in the PITX1 mRNA by the specificity domain of the helicase RHAU. PLoS One 10(12):e0144510

112. Suhl JA, Chopra P, Anderson BR, Bassell GJ, Warren ST (2014) Analysis of FMRP mRNA target datasets reveals highly associated mRNAs mediated by G-quadruplex structures formed via clustered WGGA sequences. Hum Mol Genet 23:5479–5491

113. Zhang S, Wu Y, Zhang W (2014) G-quadruplex structures and their interaction diversity with ligands. ChemMedChem 9:899–911

114. Xiong Y-X, Huang Z-S, Tan J-H (2015) Targeting G-quadruplex nucleic acids with hetero-cyclic alkaloids and their derivatives. Eur J Med Chem 97:538–551

115. Cammas A, Dubrac A, Morel B, Lamaa A, Touriol C, Teulade-Fichou M-P, Prats H, Millevoi S (2015) Stabilization of the G-quadruplex at the VEGF IRES represses cap-independent translation. RNA Biol 12(3):320–329

116. Takahama K, Miyawaki A, Shitara T, Mitsuya K, Morikawa M, Hagihara M, Kino K, Yamamoto A, Oyoshi T (2015) G-Quadruplex DNA- and RNA-specific-binding proteins engineered from the RGG domain of TLS/FUS. ACS Chem Biol 10(11):2564–2569

117. Khersonsky O, Fleishman SJ (2016) Why reinvent the wheel? Building new proteins based on ready-made parts. Protein Sci 25:1179–1187

118. Sharma VK, Sharma RK, Singh SK (2014) Antisense oligonucleotides: modifications and clinical trials. Med Chem Commun 5(10):1454–1471

119. Rouleau SG, Beaudoin J-D, Bisaillon M, Perreault J-P (2015) Small antisense oligonucleo-tides against G-quadruplexes: specific mRNA translational switches. Nucleic Acids Res 43(1):595–606

120. Oyaghire SN, Cherubim CJ, Telmer CA, Martinez JA, Bruchez MP, Armitage BA (2016) RNA G-quadruplex invasion and translation inhibition by antisense γPNA oligomers. Bio-chemistry 55:1977–1988

121. Ito K, Go S, Komiyama M, Xu Y (2011) Inhibition of translation by small RNA-stabilized mRNA structures in human cells. J Am Chem Soc 133(47):19153–19159

122. Hagihara M, Yamauchi L, Seo A, Yoneda K, Senda M, Nakatani K (2010) Antisense-induced guanine quadruplexes inhibit reverse transcription by HIV-1 reverse transcriptase. J Am Chem Soc 132(32):11171–11178

123. Marin VL, Armitage BA (2005) RNA guanine quadruplex invasion by complementary and homologous PNA probes. J Am Chem Soc 127(22):8032–8033

124. Ming X, Liang B (2015) Bioconjugates for targeted delivery of therapeutic oligonucleotides. Adv Drug Deliv Rev 87:81–89

125. Liu H, Tu Z, Feng F, Shi H, Chen K, Xu X (2015) Virosome, a hybrid vehicle for efficient and safe drug delivery and its emerging application in cancer treatment. Acta Pharma 65:105–116

126. Li L, Wei Y, Gong C (2015) Polymeric nanocarriers for non-viral gene deliver. J Biomed Nanotechnol 11:739–770

127. Ezzati Nazhad Dolatabadi J, Valizadeh H, Hamishehkar H (2015) Solid lipid nanoparticles as efficient drug and gene delivery systems: recent breakthroughs. Adv Pharm Bull 5(2): 151–159

128. Yangyuoru PM, Di Antonio M, Ghimire C, Biffi G, Balasubramanian S, Mao H (2015) Dual binding of an antibody and a small molecule increases the stability of TERRA G-quadruplex. Angew Chem 127(3):924–927

129. Brosseau J-P, Lucier J-F, Lamarche A-A, Shkreta L, Gendron D, Lapointe E, Thibault P, Paquet E, Perreault J-P, Abou Elela S, Chabot B (2014) Redirecting splicing with bifunc-tional oligonucleotides. Nucleic Acids Res 42:e40

130. Dolgosheina EV, Jeng SCY, Panchapakesan SSS, Cojocaru R, Chen PSK, Wilson PD, Hawkins N, Wiggins PA, Unrau PJ (2014) RNA mango aptamer-fluorophore: a bright, high-affinity complex for RNA labeling and tracking. ACS Chem Biol 9(10):2412–2420
131. Renaud de la Faverie A, Guédin A, Bedrat A, Yatsunyk LA, Mergny J-L (2014) Thioflavin T as a fluorescence light-up probe for G4 formation. Nucleic Acids Res 42:e65
132. Gatto B, Palumbo M, Sissi C (2008) Nucleic acid aptamers based on the G-quadruplex structure: therapeutic and diagnostic potential. Curr Med Chem 16:1248–1265

Adv Biochem Eng Biotechnol (2020) 170: 21–36
DOI: 10.1007/10_2017_21
© Springer International Publishing AG 2017
Published online: 4 August 2017

Challenges and Perspectives in Nucleic Acid Enzyme Engineering

Darko Balke, Robert Hieronymus, and Sabine Müller

Abstract Engineering of nucleic acids has been a goal in research for many years. Since the discovery of catalytic nucleic acids (ribozymes and DNAzymes), this field has attracted even more attention. One reason for the increased interest is that a large number of ribozymes have been engineered that catalyze a broad range of reactions of relevance to the origin of life. Another reason is that the structures of ribozymes or DNAzymes have been modulated such that activity is dependent on allosteric regulation by an external cofactor. Such constructs have great potential for application as biosensors in medicinal or environmental diagnostics, and as molecular tools for control of cellular processes. In addition to the development of nucleic acid enzymes by in vitro selection, rational design is a powerful strategy for the engineering of ribozymes or DNAzymes with tailored features. The structures and mechanisms of a large number of nucleic acid catalysts are now well understood. Therefore, specific design of their functional properties by structural modulation is a good option for the development of custom-made molecular tools. For rational design, several parameters have to be considered, and a number of tools are available to help/guide sequence design. Here, we discuss sequence, structural and functional design using the example of hairpin ribozyme variants to highlight the challenges and opportunities of rational nucleic enzyme engineering.

Keywords Cleavage, Engineering, Ligation, Recombination, Ribozyme

---

D. Balke, R. Hieronymus, and S. Müller (✉)
Ernst-Moritz-Arndt-Universität Greifswald, Institut für Biochemie, Felix-Hausdorff Str. 4,
17487 Greifswald, Germany
e-mail: smueller@uni-greifswald.de

Contents

1 Introduction

Over the past decade, the engineering of nucleic acid enzymes has become a powerful area of research with potential applications in the fields of chemical and molecular biology and medicinal and environmental diagnostics [1]. Ribozymes and DNAzymes are versatile molecular tools and their relevance for the aforementioned research fields has constantly grown over the past few years. Ribozyme applications in molecular biology range from simple cleavage or ligation of a defined RNA target [2], to the introduction of sequence alterations and/or modifications of the desired target RNA, to regulation of gene expression when combined with a suitable sensor module (e.g., an aptamer) [3, 4].

There are two major strategies for nucleic acid enzyme engineering: (1) in vitro evolution, which is based on selection of a nucleic acid molecule with desired properties from a library of random sequences, and (2) rational design, which starts from a known ribozyme or DNAzyme and is based on structural manipulation to affect the function in a predefined way. In vitro evolution has allowed development of many nucleic acid enzymes with novel activities and, thus, greatly enlarged the repertoire of nucleic acid catalysis [5, 6], whereas rational design has been more focused on using the intrinsic catalytic features of ribozymes and DNAzymes for novel developments. For the latter, deep knowledge of the structure and mechanism is of utmost importance. Over the years, an enormous amount of data has been collected on the structure and mechanism of nucleic acid enzymes [2, 7]. We now understand many RNA- and DNA-based catalysts well enough to turn them into useful tools. The past decade has seen impressive developments based on the usage of known catalytic nucleic acid structures [8]. For example, self-splicing group I introns have been designed to support RNA circularization [9, 10]; several ribozymes and DNAzymes have been engineered for regulation by allosteric cofactors or temperature [3]; and hairpin ribozyme descendants have been designed to support RNA repair, recombination, oligomerization, and circularization [11]. Looking into the literature, it is fascinating to see how well the engineered nucleic acid catalysts perform the intended action. However, often it takes a long time and much effort to reach that point.

Many aspects need to be considered when designing even an already extensively characterized ribozyme for a novel application. Many hurdles and challenges, including sequence design, site-specificity, structural design, and target accessibility, need to be overcome. Therefore, the design of a new ribozyme-based application

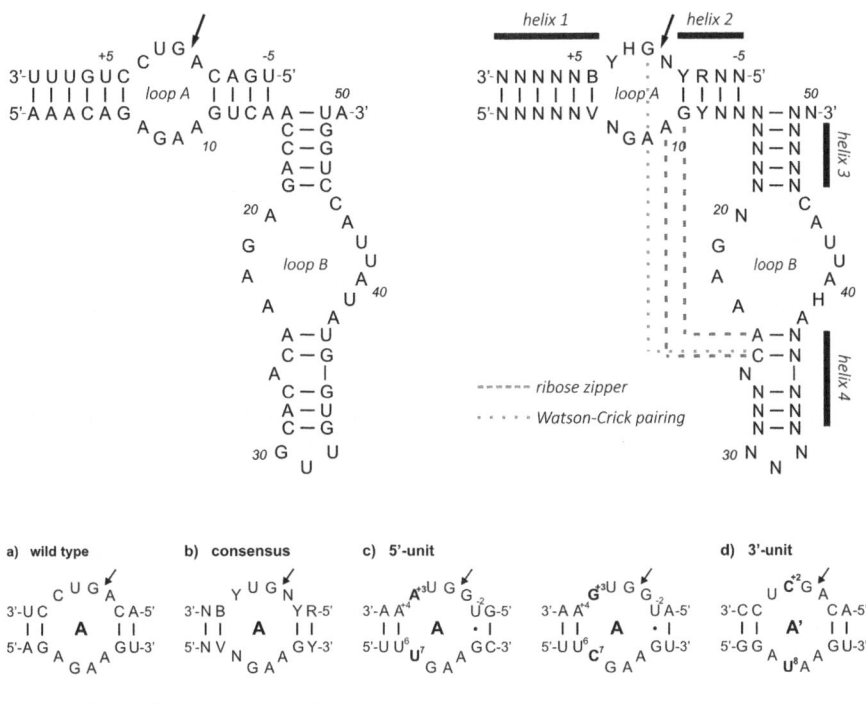

Fig. 1 Secondary structure of the hairpin ribozyme (*left*) and conserved nucleobases (*right*) with tertiary contacts. (**a–c**) Active sequence variants of the loop A motif

requires good guidelines to achieve a functional system. Design can be divided into three major parts. The first part covers sequence design, the second part deals with structural aspects that need to be taken into account for a certain application, and, third, functional design into novel activities plays a role. In this chapter, we concentrate on ribozymes (not DNAzymes) in our discussion of the challenges of rational design. In particular, we focus on the hairpin ribozyme (Fig. 1), because it is a well-studied naturally occurring RNA that we have used in our laboratory for a number of engineering projects.

2 Sequence Design

Most naturally occurring ribozymes catalyze similar reactions, which are cleavage and/or ligation of phosphodiester bonds by transesterification or hydrolysis. These activities are required for applications in molecular biology and medicine, such as specific cleavage of a defined target RNA or joining two RNA fragments. However, the ribozyme sequence needs to be adapted to bind the chosen substrate, and one has to decide which ribozyme is the most suitable for the intended ribozyme-based

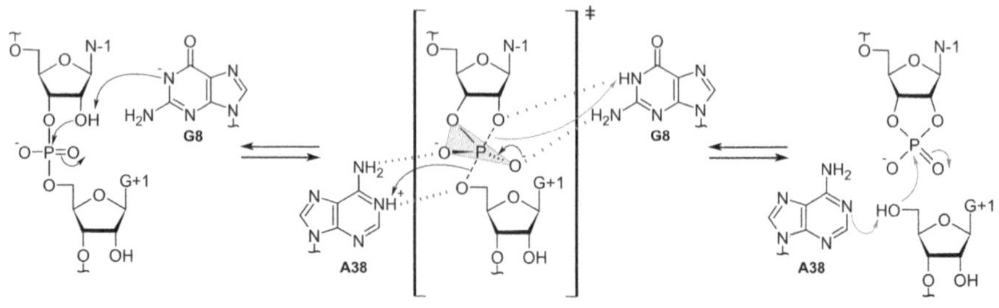

Fig. 2 Mechanism of the reversible cleavage of a phosphodiester bond catalyzed by the hairpin ribozyme

application. The hammerhead, hairpin, and hepatitis delta virus ribozymes are excellent for RNA cleavage and are useful tools for processes such as knocking down gene expression by cleaving a target mRNA [12, 13]. Group I ribozymes and hairpin ribozymes can be used for RNA sequence alteration [14–16] or, among other ribozyme motifs, for the introduction of modifications into the desired RNA strand [17, 18]. The hairpin ribozyme is employable in various ways because of its flexible adaptability to a desired target, activity, and application. In addition, this ribozyme has been extensively studied over the past decades. The three-dimensional structure has been solved and the reaction mechanism is well understood [19] (Fig. 2).

A large part of the hairpin ribozyme sequence consists of variable nucleotides, which makes it relatively easy to tailor the ribozyme for specific RNA targets. Furthermore, with a length of 50 nucleotides (nt), the minimal structure of the hairpin ribozyme (a *trans*-acting ribozyme) represents a relatively small catalytic RNA, which is easy to handle (low tendency to misfold) and synthesize. The minimal hairpin ribozyme consists of four base-paired helices (H1–H4) and two loops (A and B) (Fig. 1). The cleavage/ligation site is located in loop A. The active conformation is formed by docking of loops A and B. Interestingly, the helical junction has a tremendous effect on the stability of the docked conformer [20]. Four-way junctions provide a stable scaffold that, in the case of the hairpin ribozyme, enables stabilization of the tertiary structure and thus promotes ligation [21] (Fig. 3). A hairpin ribozyme with a four-way junction binds its cleavage product with higher affinity than the minimal hairpin motif does, because tertiary interactions within the folded structure contribute to product binding. Two-way and three-way junctions are less stable, but more sensitive to regulation by ligands [20]. The crystal structure of the hairpin ribozyme was solved by Rupert and Ferré d'Amaré and gives insight into the catalytic mechanism, which is thought to proceed by general acid–base catalysis [22, 23].

The most crucial aspect for sequence design is the consensus sequence of the ribozyme, which defines the ribozyme's adaptability to a particular target RNA. For the hairpin ribozyme, the conserved nucleobases essential for formation of the catalytically active structure and for active site chemistry are only located within loop A and loop B [24–28]. The helical regions are fully variable and can be easily

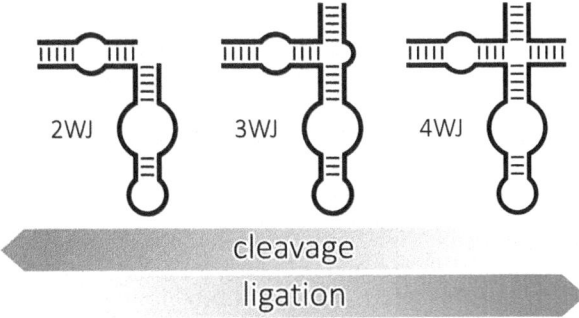

Fig. 3 Influence of the hairpin ribozyme structure containing a two-way (*2WJ*), three-way (*3WJ*), or four-way (*4WJ*) junction on cleavage/ligation activity

adapted to the target RNA. For a *trans*-acting ribozyme, loop A is formed upon binding of the substrate to the ribozyme. Therefore, it is important to screen the target RNA for the required consensus sequence $5'-Y_{-2} \ N_{-1}{\downarrow}G_{+1}U_{+2}Y_{+3}B_{+4}-3'$ (with N = A, C, G, or U; Y = C or U; B = C, G, or U) to ensure excellent ribozyme activity (Fig. 1b). However, deviations from the consensus sequence do not necessarily result in loss of ribozyme activity. Although the presence of G+1 is indispensable [24, 26, 29, 30], other deviations are more tolerated. As shown previously, A+4, although not allowed according to the consensus sequence mentioned above, does not lead to a significant decrease in ribozyme activity; furthermore, loss of activity caused by deviations from U+2 and Y+3 in the substrate strand can be restored by compensatory mutations in the ribozyme strand [31–33] (Fig. 1c).

Compensatory mutations can be found by careful checking of the hairpin ribozyme crystal structure and by trial and error activity tests. This strategy requires some effort; however, it has allowed the design of ribozymes and processing of substrates beyond the consensus sequence. Interestingly, there are also mutations that strongly influence the cleavage–ligation equilibrium of the hairpin ribozyme. In the wild type, ribozyme ligation is favored over cleavage, but mutations of A9 and A10 eliminate ligation activity and leave cleavage activity fully intact [34]. Moreover, mutation of A10→G enhances cleavage fivefold but prevents ligation, and substrates with A10→C are ligated but virtually uncleaved [32]. Thus, a single point mutation can have both quantitative and qualitative effects on activity and can be of great importance in rational design.

Another important aspect of sequence design is the length of the duplex formed between ribozyme and substrate upon binding, as it can strongly influence the preference for cleavage or ligation catalysis depending on the stability of the ribozyme–substrate/product complex. When the duplex is relatively short, resulting in a less stable ribozyme–substrate complex (but stable enough to form a catalytically competent structure), then dissociation of cleavage products (with a length of about 5–8 nt) is fast, and cleavage is favored over ligation. For this reason, the equilibrium is shifted toward cleavage in minimal hairpin ribozymes consisting of just two hinged loop A and loop B domains. On the other hand, a longer duplex leads to tighter bound substrates because of the large thermodynamic contribution of the Watson–Crick base pairs. When fragments (with a length of ≥ 10 nt) are tightly bound to the ribozyme, dissociation is not favored and the hairpin ribozyme preferentially undergoes ligation. Increased stability of the ribozyme–substrate complex can be achieved by lengthening the $3'$-end of the ribozyme via a three-way junction (Fig. 3).

Three-way junction hairpin ribozymes can be used for RNA ligation. Thereby, two RNA substrates can be joined to form a long-mer RNA, which would be inaccessible by chemical synthesis. The ability to produce long-mer RNAs becomes even more important when modifications are site-specifically introduced. The RNA fragment that contains the desired modification (e.g., fluorescent dye or biotin) can be chemically synthesized and subsequently ligated to a second RNA fragment in a reaction supported by the three-way junction hairpin ribozyme. Three-way junction hairpin ribozymes can also be used for RNA recombination. When combining two three-way junction hairpin ribozymes into one molecule, a new type of catalytic RNA is generated, which we have named "twin ribozyme" (Fig. 4).

Twin ribozymes are capable of cleavage and ligation of a suitable RNA substrate at two defined positions, allowing the exchange of a short patch of RNA for an externally added oligonucleotide [35]. The twin ribozyme-mediated exchange reaction enables sequence alteration or introduction of modifications into the target RNA. However, because binding of the externally given oligonucleotide competes with re-association of the internal cleavage fragment, the equilibrium needs to be shifted toward binding of the external oligonucleotide to ensure optimal sequence exchange. This shift can be achieved by promoting dissociation of the internal cleavage fragment. Therefore, binding of the substrate RNA is designed such that a destabilizing structure (e.g., mismatch or bulge) is formed within the sequence patch to be cut out. Consequently, dissociation of the cleavage fragment is promoted. The externally added oligonucleotide forms a contiguous duplex with the ribozyme and is preferentially bound because of its more stable and, therefore, favored structure (Fig. 4). Finally, after twin ribozyme-mediated ligation, the desired product is formed. Depending on the design of the ribozyme–substrate complex, twin ribozymes can mediate the exchange of fragments of the same length [18] or exchange of short fragments by longer versions [35] and vice versa [36]. A crucial aspect for optimal twin ribozyme-mediated sequence exchange is the length of the fragment to be cut out. It is very important that the gap between the two cleavage/ligation sites is not too large, otherwise dissociation of the cut-out fragment and, consequently, exchange with the repair oligonucleotide is dramatically hampered. As a guideline, the optimal length of the fragment to be cut out should be 12–18 nt to ensure sufficient dissociation [18, 31, 32, 35, 36].

Thus, a number of points have to be considered when designing the sequence of a ribozyme for a certain target and application. First, a suitable naturally occurring or previously in vitro selected nucleic acid enzyme has to be defined as the precursor or starting point for design. Next, the sequence of the ribozyme/

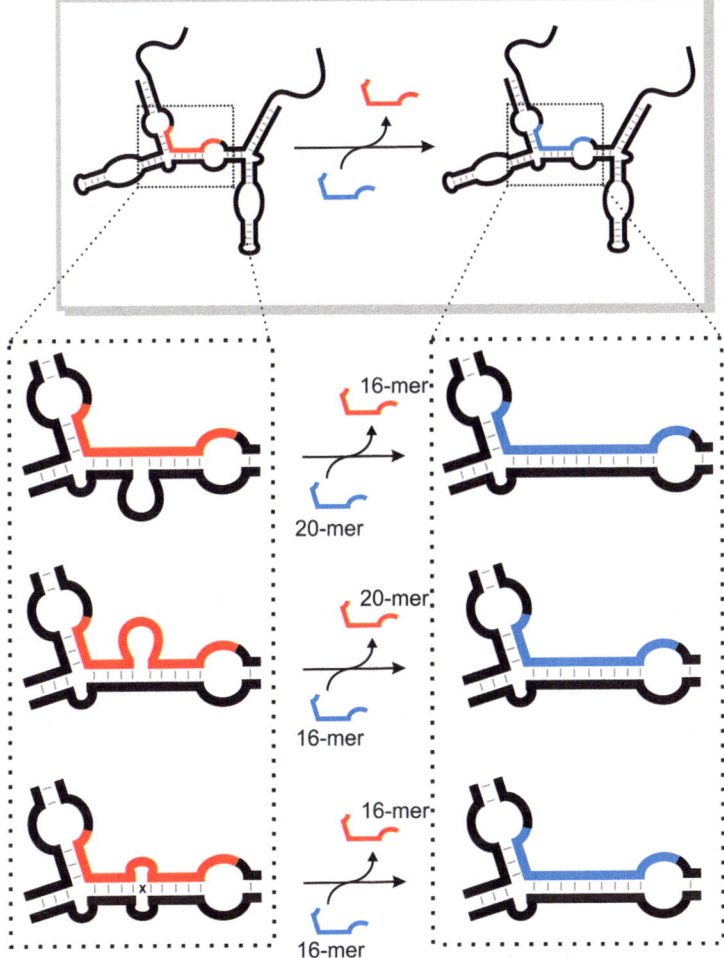

Fig. 4 Twin ribozyme-mediated fragment exchange reactions. The *red* fragment is cut out (cleavage is favored) and replaced with the *blue* fragment (ligation is favored). To shift the equilibrium toward product formation, substrate binding is designed to lead to the formation of a destabilizing structure within the sequence patch to be cut out (e.g., mismatches or bulges). In this case, dissociation of the formed cleavage fragment is promoted. In contrast, the externally added oligonucleotide forms a contiguous duplex with the ribozyme and undergoes preferential ligation because of its favorable stable structure. Upon twin ribozyme-mediated ligation, the desired product is formed

DNAzyme that is involved in substrate binding needs to be adapted to recognize and process the chosen target. This requires ensuring that sequence changes do not inhibit the activity of the nucleic acid enzyme. Last, because sequence changes can affect the reaction equilibrium (e.g., between cleavage and ligation), they need to be considered (or maybe even used on purpose) to favor one or the other activity.

These key aspects apply to all engineering work, independent of the specific ribozyme/DNAzyme and application.

3   Structural Design

To develop a rationally designed system, it is essential to verify that the engineered nucleic acid sequence folds into the intended secondary structure in the presence of its substrate. Although the helical parts of most ribozymes and DNAzymes are freely selectable, it is possible that the designed sequence folds into an energetically preferred secondary structure that is different from the structure of the active state, thus forming an inactive RNA or DNA. In addition, sequence changes necessary to meet the expected application can result in unwanted interactions. Such challenges include targets that do not bind to the substrate binding site but instead bind to the ribozyme sequence at another site, strands that favor monomolecular over bimolecular folding, and a thermodynamically favored dimer instead of an intramolecularly folded nucleic acid strand. For every substituted, inserted, or deleted nucleotide, the secondary structure of the overall system has to be rechecked to ensure that it still folds into the intended active conformation. If not, the mutation has to be reversed or (more challenging) compensatory mutations found and inserted.

A first indication of proper folding can be achieved with computer-aided folding algorithms, which focus on RNA folding but can also be applied to predict DNA folding. For prediction of RNA secondary structure, several software applications have been developed and are freely accessible; examples include RNAstructure [37], Vienna RNA Package [38, 39], and Mfold [40] (Table 1). The most popular method for predicting RNA secondary structure is based on calculating the minimal free energy of structural motifs that are formed by base-pairing within the RNA. The Gibbs free energy change can be determined by summing the individual base-paring energies. The secondary structure with the lowest Gibbs free energy change is generally the preferred structure. However, these programs only calculate Watson–Crick and wobble base pairs (G–U), and do not consider noncanonical base pairs such as Hoogsten base pairs. Pseudoknots are also usually ignored in order to gain higher calculation efficiency. Furthermore, because secondary structure prediction is a modeling approach, the calculated structure with the lowest free energy does not necessarily correspond to the actual secondary structure formed under the chosen reaction conditions. This crucial aspect should always be taken into account when using computational methods for predicting RNA secondary structure [41]. The method works very well for short and simple structures such as the hairpin ribozyme. However, the method is not accurate for large RNAs, as exemplified by the fact that only 50% of the base pairs of *Escherichia coli* 16S rRNA were predicted correctly [42].

Table 1 Useful programs for sequence design of ribozymes

Program	Description	URL
RNAstructure	RNA secondary structure prediction and prediction of the consensus secondary structure of two or more sequences (Dynalign or Multialign)	http://rna.urmc.rochester.edu/rnastructure.html
Vienna RNA Package	RNA secondary structure prediction (RNAfold) and RNA sequence design using constraint secondary structures (RNAinverse)	http://www.tbi.univie.ac.at/RNA/
RNAshapes	Secondary structure prediction of multiple sequences, followed by determination of a conserved structure	http://bibiserv.techfak.uni-bielefeld.de/rnashapes
RNA Designer	RNA sequence design using constraint secondary structures	http://www.rnasoft.ca/cgi-bin/RNAsoft/RNAdesigner/rnadesign.pl
RNAdesign	RNA design with multiple target secondary structures	http://www.bioinf.uni-leipzig.de/~choener/rnadesign/
VARNA	Visualization and drawing of RNA secondary structures	http://varna.lri.fr/

Refinement of the secondary structure prediction can be achieved by incorporating experimental information into the prediction algorithm [43]. Thus, prediction of the 16S RNA secondary structure was improved to 72% accuracy by including experimental data obtained from chemical probing experiments, and up to 95% accuracy by selective 2′-hydroxyl acylation analyzed by primer extension (SHAPE) [42]. Therefore, one should keep in mind that computer-aided prediction of secondary structure works best for smaller RNAs. For larger RNAs, the accuracy of prediction is significantly improved by additional experimental data, most favorably from SHAPE analysis. The software platform's RNA structure offers the possibility of directly feeding in experimental SHAPE data, which then are considered in the structure calculation. The emerging abundance of experimental data not only helps to refine prediction of secondary structure when using current folding algorithms, but also to improve the prediction algorithms or to develop new, more accurate algorithms.

As an alternative to verifying the secondary structure of a designed RNA, one can apply a method that allows inverse RNA sequence design. In contrast to the abovementioned approach, sequence design proceeds in the opposite way. First, the desired secondary structure is defined and then an inverse RNA sequence design program, such as RNAinverse (included in the Vienna RNA package [44]) or RNA Designer [45], determines the RNA sequence with the lowest free energy that gives the predefined secondary structure. Because the entire RNA sequence may not be variable, it is possible to specify nucleotides at defined positions within the secondary structure. More recently, several inverse folding programs (e.g., MODENA) have been developed that even allow the design of RNA sequences that fold into multiple target secondary structures [46, 47]. This tool could be very useful for the design of riboswitches, aptazymes, or multiple substrate-processing ribozymes.

It should be mentioned that all efforts in ribozyme design are useless if the ribozyme binding site within the target RNA is not accessible. This applies not only to ribozyme design, but also to antisense oligonucleotides (ASOs), short interfering RNAs (siRNAs), and guide-RNAs that direct a specific enzyme to the desired processing location. As previously shown, a twin ribozyme was developed by rational design and was able to repair a three-base deletion within a short model substrate based on *CTNNB1* mRNA with a yield of 30% [32]. However, after adaption of the twin ribozyme to the entire *CTNNB1*-ΔS45 mRNA repair, the reaction failed. SHAPE analysis to refine secondary structure prediction revealed that the mRNA folded into an unfavorable structure, such that the twin ribozyme-binding site was blocked. To overcome that challenge one can follow several approaches. A simple and sometimes very helpful technique is the usage of competitor oligonucleotides that assist in defolding the cleavage/ligation of the target RNA [18]. A more systematic approach for detection of ribozyme binding sites deals with preparation of an oligodeoxynucleotide (ODN) library used for an RNaseH assay [48]. Effective cleavage of the RNA–DNA hybrid by RNaseH marks the most accessible sites for ribozyme base-pairing. Another sophisticated technique makes use of RNA–protein hybrid ribozymes that are able to process any RNA target independently of the secondary or tertiary structure [49]. To do so, the constitutive transport element (CTE), an RNA motif that allows interaction with intracellular RNA helicases, has to be conjugated to the ribozyme terminus. The bound RNA helicase assists the ribozyme to bind its target site by unwinding the local secondary structure.

4 Functional Design

As mentioned above, in many of our applications we took advantage of the cleavage/ligation equilibrium of the hairpin ribozyme, which can be easily shifted via temperature adjustment and/or substrate stabilization (or destabilization), thus allowing control of the two reactions. Using the example of a hairpin ribozyme-mediated recombination system (Fig. 5), we discuss the challenges of applying this control for functional design.

As a key feature of the engineered recombination system, two RNA strands without function are cleaved, mutually exchanged, and recombined by a single hairpin ribozyme to yield a functional RNA [14]. The two RNA educts each consist of a nonsense half and a profunctional half linked via the hairpin ribozyme specific cleavage sequence AGUC. The nonsense part was designed such that it binds weakly to the substrate binding site of the ribozyme, whereas the profunctional part binds strongly. Upon cleavage, dissociation of the nonsense part is facilitated, resulting in preferential cleavage without significant back-ligation. Because of the low binding affinity of the nonsense fragment (but strong enough that cleavage can

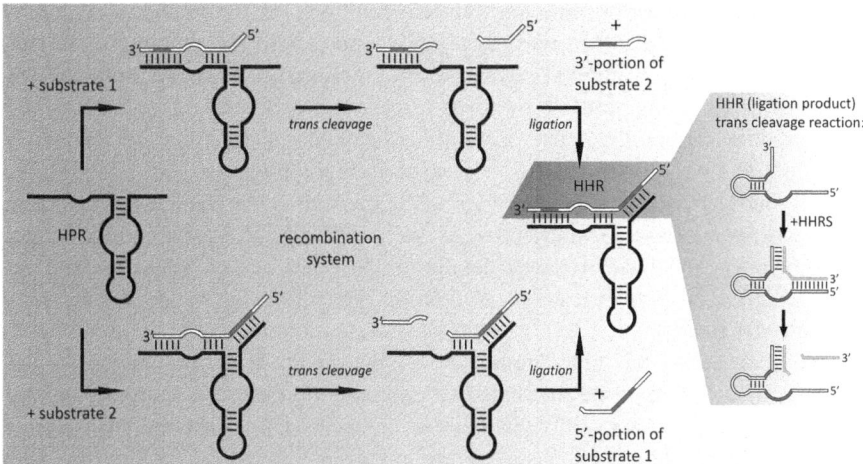

Fig. 5 RNA recombination mediated by the hairpin ribozyme (*HPR*). The recombination product, a functional hammerhead ribozyme (*HHR*), is able to catalyze cleavage of an externally added substrate (*HHRS*), which is depicted in *light grey*. The conserved nucleotide regions of HHR are shown in *dark grey*

occur), exchange between the nonsense and profunctional fragments at the binding site is promoted. However, exchange requires that the profunctional fragments do not bind too tightly so that dissociation and association can occur next to each other. Once both profunctional fragments are bound to one ribozyme, ligation is the favored reaction because of the strong binding of both fragments and the very slow dissociation, such that an active recombination product emerges.

The engineering of this recombination system was challenging because a single ribozyme had to bind two different RNA substrates equally well and catalyze cleavage, releasing the nonsense part and recombining the two profunctional parts. This required finding a good balance between dissociation of one of the profunctional parts from one ribozyme molecule and re-association with another ribozyme molecule for ligation to the profunctional part located there. Thus, a key aspect of design was tuning the lengths of individual helices of ribozyme–substrate complexes. According to the hairpin ribozyme consensus structure, helix 1 needs to consist of four base pairs or longer, and helix 2 is required to have exactly four base pairs. This restriction made it difficult to create a discriminating binding site for one of the substrates. In addition, the four-base pair helix 2 is presumably too weak for the profunctional part to undergo proper ligation. Therefore, we integrated a fifth helix, thus creating a three-way junction hairpin ribozyme (compare Fig. 3); only the substrate with the profunctional part at the 5′-end is bound to this three-way junction. As a side effect, the additional base pairs formed in helix 5 not only make dissociation of the required fragment less favorable, but also enhance overall ribozyme efficiency by facilitating formation of the active conformation [50]. Because the lengths of helices 1 and 5 are not limited, extension of the duplex formed between the ribozyme and the two profunctional fragments can increase ligation tremendously. However, this would also add to the problem of product inhibition. Strongly hampered product dissociation makes fragment exchange less probable and inhibits multiple turnover reactions. Furthermore, if the recombination product cannot dissociate from the ribozyme, its functionality cannot be exploited and a successful reaction cascade to recombination cannot be verified. Taking into account all these considerations, one can conclude that the lengths of helices between substrates and ribozyme, and thus the binding affinity of substrates, intermediates, and final products, determine the efficiency of fragment exchange and the preference for cleavage or ligation. Therefore, helix length has to be carefully adjusted.

We approached this challenging part of the design in a retro-synthetic way. Starting with design of the functional recombination product (a *trans*-acting hammerhead ribozyme, also quite variable in sequence), we inserted the required AGUC sequence into a nonconserved region for cleavage by the hairpin ribozyme. The ribozyme binding domain was designed for optimal binding of the recombination product with minimal structural distortion of the overall system. We continued by designing the educts (the two RNA strands without function at the beginning of the reaction chain) on the basis of the prior defined binding site. Because the nonsense part of one substrate had to bind less strongly than its profunctional counterpart in the other substrate, we took into account considerations such as mismatches, GU wobble pairs, and shortening of helix length. Nevertheless, it was necessary to achieve a binding capacity that was strong enough to ensure formation of a catalytically competent structure and substrate cleavage (4 bp in helices 1 and 2). Furthermore, undesired interactions of the ribozyme or substrates with the RNAs in the system had to be limited to an insignificant amount. If the defined preconditions could not be met by the designed structure, we went back to the recombination product and the substrate binding site, and re-designed their sequence and length, if necessary going through iterative cycles of design and theoretical verification. Because recombination should proceed in a one-pot reaction, the two initial ribozyme–substrate complexes were designed to have free energy values as close as possible to each other, and to cleave both substrates efficiently enough to deliver sufficient amounts of profunctional fragments for recombination. This required further adjustment of the designed sequences, but still taking into account all the aforementioned conditions. At the end of the design process, we successfully engineered a hairpin ribozyme-based recombination system, composed of two substrates and one ribozyme (Fig. 5), that performed recombination with high yield [14].

5 Summary

Engineering of nucleic acid catalysts by rational design is a powerful tool with potential applications. However, it can be a very challenging task, requiring attention to be paid to a number of factors associated with sequence adaptation, folding and active conformation, and reaction equilibria. Usually, engineering starts from a precursor ribozyme or DNAzyme with known catalytic features. As discussed, one of the key steps is adaptation of the ribozyme sequence to recognize a defined target, ensuring that activity is undamaged. Substrate association and dissociation processes can be influenced by variations in the lengths of substrate binding domains, but the ribozyme/DNAzyme must still be able to fold into the required active conformation. Furthermore, structural modulation of the chosen precursor enzyme can be used to influence the reaction equilibrium (e.g., for the hairpin ribozyme, structural stabilization favors ligation, whereas destabilization favors cleavage).

A number of tools are available to aid rational design, in particular software and platforms for theoretical prediction of nucleic acid secondary structures that can help guide sequence design. One of the most important prerequisites for engineering is that the structure and mechanistic properties of the starting nucleic acid are known. The more data are available, the higher is the chance for successful design. Nowadays, this prerequisite is met by a number of nucleic acids. Many of the ribozymes and DNAzymes known today are understood to a level that allows them to be turned into useful tools.

References

1. Liu J, Cao Z, Lu Y (2009) Functional nucleic acid sensors. Chem Rev 109(5):1948–1998. doi:10.1021/cr030183i
2. Jimenez RM, Polanco JA, Luptak A (2015) Chemistry and biology of self-cleaving ribozymes. Trends Biochem Sci 40(11):648–661. doi:10.1016/j.tibs.2015.09.001
3. Frommer J, Appel B, Müller S (2015) Ribozymes that can be regulated by external stimuli. Curr Opin Biotechnol 31:35–41. doi:10.1016/j.copbio.2014.07.009
4. Müller S, Appel B, Krellenberg T, Petkovic S (2012) The many faces of the hairpin ribozyme: structural and functional variants of a small catalytic RNA. IUBMB Life 64(1):36–47. doi:10.1002/iub.575
5. Hollenstein M (2015) DNA catalysis: the chemical repertoire of DNAzymes. Molecules 20(11):20777–20804. doi:10.3390/molecules201119730
6. Jäschke A (2001) Artificial ribozymes and deoxyribozymes. Curr Opin Struct Biol 11(3):321–326
7. Wilson TJ, Lilley DM (2015) RNA catalysis—is that it? RNA 21(4):534–537. doi:10.1261/rna.049874.115
8. Müller S, Appel B, Balke D, Hieronymus R, Nübel C (2016) Thirty-five years of research into ribozymes and nucleic acid catalysis: where do we stand today? F1000Res, 5. doi:10.12688/f1000research.8601.1

9. Umekage S, Kikuchi Y (2009a) In vitro and in vivo production and purification of circular RNA aptamer. J Biotechnol 139(4):265–272. doi:10.1016/j.jbiotec.2008.12.012

10. Umekage S, Kikuchi Y (2009b) In vivo circular RNA production using a constitutive promoter for high-level expression. J Biosci Bioeng 108(4):354–356. doi:10.1016/j.jbiosc.2009.04.011

11. Müller S (2015) Engineering of ribozymes with useful activities in the ancient RNA world. Ann N Y Acad Sci 1341:54–60. doi:10.1111/nyas.12695

12. Burnett JC, Rossi JJ (2012) RNA-based therapeutics: current progress and future prospects. Chem Biol 19(1):60–71. doi:10.1016/j.chembiol.2011.12.008

13. Drude I, Dombos V, Vauléon S, Müller S (2007a) Drugs made of RNA: development and application of engineered RNAs for gene therapy. Mini Rev Med Chem 7(9):912–931

14. Hieronymus R, Godehard SP, Balke D, Müller S (2016) Hairpin ribozyme mediated RNA recombination. Chem Commun 52:4365–4368. doi:10.1039/C6CC00383D

15. Müller S (2003) Engineered ribozymes as molecular tools for site-specific alteration of RNA sequence. Chembiochem 4(10):991–997. doi:10.1002/cbic.200300665

16. Riley CA, Lehman N (2003) Generalized RNA-directed recombination of RNA. Chem Biol 10 (12):1233–1243

17. Dotson 2nd PP, Frommeyer KN, Testa SM (2008) Ribozyme mediated trans insertion-splicing of modified oligonucleotides into RNA. Arch Biochem Biophys 478(1):81–84. doi:10.1016/j. abb.2008.07.010

18. Vauléon S, Ivanov SA, Gwiazda S, Müller S (2005) Site-specific fluorescent and affinity labelling of RNA by using a small engineered twin ribozyme. Chembiochem 6 (12):2158–2162. doi:10.1002/cbic.200500215

19. Ferre-D'Amare AR (2004) The hairpin ribozyme. Biopolymers 73(1):71–78. doi:10.1002/bip. 10516

20. Klostermeier D, Millar DP (2000) Helical junctions as determinants for RNA folding: origin of tertiary structure stability of the hairpin ribozyme. Biochemistry 39(42):12970–12978

21. Fedor MJ (1999) Tertiary structure stabilization promotes hairpin ribozyme ligation. Biochemistry 38(34):11040–11050. doi:10.1021/bi991069q

22. Rupert PB, Ferré-D'Amaré AR (2001) Crystal structure of a hairpin ribozyme-inhibitor complex with implications for catalysis. Nature 410(6830):780–786. doi:10.1038/35071009

23. Rupert PB, Massey AP, Sigurdsson ST, Ferré-D'Amaré AR (2002) Transition state stabilization by a catalytic RNA. Science 298(5597):1421–1424. doi:10.1126/science.1076093

24. Berzal-Herranz A, Joseph S, Burke JM (1992) In vitro selection of active hairpin ribozymes by sequential RNA-catalyzed cleavage and ligation reactions. Genes Dev 6(1):129–134. doi:10. 1101/gad.6.1.129

25. Berzal-Herranz A, Joseph S, Chowrira BM, Butcher SE, Burke JM (1993) Essential nucleotide sequences and secondary structure elements of the hairpin ribozyme. EMBO J 12 (6):2567–2573

26. Chowrira BM, Berzal-Herranz A, Burke JM (1991) Novel guanosine requirement for catalysis by the hairpin ribozyme. Nature 354(6351):320–322. doi:10.1038/354320a0

27. Chowrira BM, Burke JM (1991) Binding and cleavage of nucleic acids by the "hairpin" ribozyme. Biochemistry 30(35):8518–8522

28. Joseph S, Berzal-Herranz A, Chowrira BM, Butcher SE, Burke JM (1993) Substrate selection rules for the hairpin ribozyme determined by in vitro selection, mutation, and analysis of mismatched substrates. Genes Dev 7(1):130–138. doi:10.1101/gad.7.1.130

29. Anderson P, Monforte J, Tritz R, Nesbitt S, Hearst J, Hampel A (1994) Mutagenesis of the hairpin ribozyme. Nucleic Acids Res 22(6):1096–1100. doi:10.1093/nar/22.6.1096

30. Shippy R, Siwkowski A, Hampel A (1998) Mutational analysis of loops 1 and 5 of the hairpin ribozyme. Biochemistry 37(2):564–570. doi:10.1021/bi9721288

31. Balke D, Becker A, Müller S (2016) In vitro repair of a defective EGFP transcript and translation into a functional protein. Org Biomol Chem 14:6729–6737. doi:10.1039/ c6ob01043a

32. Balke D, Zieten I, Strahl A, Müller O, Müller S (2014) Design and characterization of a twin ribozyme for potential repair of a deletion mutation within the oncogenic *CTNNB1*-ΔS45 mRNA. ChemMedChem 9(9):2128–2137. doi:10.1002/cmdc.201402166

33. Drude I, Strahl A, Galla D, Müller O, Müller S (2011) Design of hairpin ribozyme variants with improved activity for poorly processed substrates. FEBS J 278(4):622–633. doi:10.1111/j.1742-4658.2010.07983.x

34. Gaur S, Heckman JE, Burke JM (2008) Mutational inhibition of ligation in the hairpin ribozyme: substitutions of conserved nucleobases A9 and A10 destabilize tertiary structure and selectively promote cleavage. RNA 14(1):55–65. doi:10.1261/rna.716108

35. Welz R, Bossmann K, Klug C, Schmidt C, Fritz HJ, Müller S (2003) Site-directed alteration of RNA sequence mediated by an engineered twin ribozyme. Angew Chem Int Ed 42 (21):2424–2427. doi:10.1002/anie.200250611

36. Drude I, Vauléon S, Müller S (2007b) Twin ribozyme mediated removal of nucleotides from an internal RNA site. Biochem Biophys Res Commun 363(1):24–29. doi:10.1016/j.bbrc.2007.08.135

37. Reuter JS, Mathews DH (2010) RNAstructure: software for RNA secondary structure prediction and analysis. BMC Bioinform 11:129. doi:10.1186/1471-2105-11-129

38. Hofacker IL (2003) Vienna RNA secondary structure server. Nucleic Acids Res 31(13):3429–3431

39. Hofacker IL, Fontana W, Stadler PF, Bonhoeffer LS, Tacker M, Schuster P (1994) Fast folding and comparison of RNA secondary structures. Monatsh Chem 125(2):167–188. doi:10.1007/Bf00818163

40. Zuker M (2003) Mfold web server for nucleic acid folding and hybridization prediction. Nucleic Acids Res 31(13):3406–3415

41. Weeks KM, Mauger DM (2011) Exploring RNA structural codes with SHAPE chemistry. Acc Chem Res 44(12):1280–1291. doi:10.1021/ar200051h

42. Deigan KE, Li TW, Mathews DH, Weeks KM (2009) Accurate SHAPE-directed RNA structure determination. Proc Natl Acad Sci U S A 106(1):97–102. doi:10.1073/pnas.0806929106

43. Mathews DH, Disney MD, Childs JL, Schroeder SJ, Zuker M, Turner DH (2004) Incorporating chemical modification constraints into a dynamic programming algorithm for prediction of RNA secondary structure. Proc Natl Acad Sci U S A 101(19):7287–7292. doi:10.1073/pnas.0401799101

44. Lorenz R, Bernhart SH, Höner zu Siederdissen C, Tafer H, Flamm C, Stadler PF, Hofacker IL (2011) ViennaRNA package 2.0. Algorithms Mol Biol 6:26. doi:10.1186/1748-7188-6-26

45. Andronescu M, Fejes AP, Hutter F, Hoos HH, Condon A (2004) A new algorithm for RNA secondary structure design. J Mol Biol 336(3):607–624. doi:10.1016/j.jmb.2003.12.041

46. Höner zu Siederdissen C, Hammer S, Abfalter I, Hofacker IL, Flamm C, Stadler PF (2013) Computational design of RNAs with complex energy landscapes. Biopolymers 99 (12):1124–1136. doi:10.1002/bip.22337

47. Taneda A (2015) Multi-objective optimization for RNA design with multiple target secondary structures. BMC bioinformatics 16:280. doi:10.1186/s12859-015-0706-x

48. Scherr M, LeBon J, Castanotto D, Cunliffe HE, Meltzer PS, Ganser A, Riggs AD, Rossi JJ (2001) Detection of antisense and ribozyme accessible sites on native mRNAs: application to *NCOA3* mRNA. Mol Ther 4(5):454–460. doi:10.1006/mthe.2001.0481

49. Warashina M, Kuwabara T, Kato Y, Sano M, Taira K (2001) RNA-protein hybrid ribozymes that efficiently cleave any mRNA independently of the structure of the target RNA. Proc Natl Acad Sci U S A 98(10):5572–5577. doi:10.1073/pnas.091411398

50. Welz R, Schmidt C, Müller S (2001) Spermine supports catalysis of hairpin ribozyme variants to differing extents. Biochem Biophys Res Commun 283(3):648–654. doi:10.1006/bbrc.2001.4829

Adv Biochem Eng Biotechnol (2020) 170: 37–58
DOI: 10.1007/10_2016_59
© Springer International Publishing AG 2017
Published online: 23 June 2017

Strategies for Characterization of Enzymatic Nucleic Acids

Fatemeh Javadi-Zarnaghi and Claudia Höbartner

Abstract Practical application of enzymatic nucleic acids has received more attention in recent years. Understanding the mechanism of catalysis and availability of information on potentials and limitations of these enzymes expands their application scope. A general approach for characterization of functional macromolecules including enzymatic nucleic acids is to perturb a specific set of condition and to follow the perturbation effect by biophysical and biochemical methods. This chapter reviews several perturbation strategies for functional nucleic acids, including deletion, mutation, and modifications of backbone and nucleobases, and consequent kinetic analysis, spectroscopic investigations, and probing assays. In addition to single point mutation and modifications, different combinatorial approaches for perturbation interference analysis provide reliable high amounts of data in a time-effective manner. The chapter compares various combinatorial perturbation interference analysis methods, that is, combinatorial mutation interference analysis (CoMA), nucleotide analogue interference mapping for RNA and DNA (NAIM and dNAIM), chemical and enzymatic combinatorial nucleoside deletion scanning (NDS), and dimethyl sulfate interference (DMSi).

Keywords CoMA dNAIM, Combinatorial perturbation interference analysis, Enzymatic NDS and DMS interference, Functional nucleic acids, NAIM, NDS

F. Javadi-Zarnaghi (✉)
Department of Biology, Faculty of Sciences, University of Isfahan, Isfahan, Iran
e-mail: Fa.javadi@sci.ui.ac.ir

C. Höbartner
Institut für Organische und Biomolekulare Chemie, Georg-August-Universität Göttingen, Göttingen, Germany

Contents

Abbreviations

CoMA   Combinatorial mutation interference analysis
DMSi   Dimethyl sulfate interference
dNAIM   Nucleotide analogue interference mapping of DNA
NAIM   Nucleotide analogue interference mapping
NDS   Combinatorial nucleoside deletion scanning

1   Enzymatic Nucleic Acids

Enzymatic nucleic acids consist of three groups of macromolecules – ribozymes [1], deoxyribozymes [2], and catalysts from synthetic genetic polymers [3]. From this list, only ribozymes have naturally present instances including hammerhead ribozyme [4], hairpin ribozyme [5], hepatitis delta virus (HDV) ribozyme [6], Varkud satellite (VS) ribozyme [7], twister ribozyme [8], glmS ribozyme [9], group I and II introns [10, 11], and RNA-dependent RNase P [12]. Besides natural ribozymes, many in vitro selections expanded the world of enzymatic nucleic acids. Artificial ribozymes, deoxyribozymes, and enzymes from genetic polymers have been selected in vitro to catalyze a variety of reactions. Examples of artificial ribozymes include class I, II, and III [13] and L1 ligase [14] ribozymes, RNA-aminoacylating ribozyme, that is, flexizyme [15], and RNA-cleaving ribozymes such as leadzyme [16]. In contrast to ribozymes, all deoxyribozymes that have been reported to date are synthetic. Many of these catalyze phosphodiester bond modifications and are RNA- and DNA-cleaving or ligating deoxyribozymes [17]. Phosphodiester bond-modifying enzymes have been utilized for practical applications. Current reports on application of these deoxyribozymes include in vivo RNA cleavage [18], post transcriptional site-specific labeling of RNAs [19], and utilization in logic gates [20] or as parts of biosensors [21]. The latter is to the extent that in vitro selection strategies have been developed to select specifically deoxyribozymes that are useful for biosensing [22].

Recently, selections for enzymatic nucleic acids that are capable of catalysis beyond phosphodiester bond-modification such as DNA-glycosylation [23], DNA phosphorylation [24], and peptide modification and cleavage [25, 26] have received more attention. Expanding the scope of nucleic acid catalysis is promising for the development of new creative practical application of such catalysts. Besides, incorporating modified nucleosides in DNA [27] and selection of catalysts from nucleic acids that bear substantial backbone modification [3] pushes the border of practical application of enzymatic nucleic acids even further.

Some application of nucleic acid enzyme may necessitate modifications and mutations. Some sequences or functional groups might be required to be added or deleted to increase stability, activity, or functionality of the enzyme. In these cases, identifying positions that tolerate perturbations is necessary. In addition, substrate specificity of enzymatic nucleic acids impacts many applications. In some applications, pH or presence of specific metal ions may be advantageous or, in contrast, may cause limitations [19]. Functional characterization of enzymatic nucleic acids provides information to adjust the conditions for improved application. In addition, functional characterization of enzymatic nucleic acids enlightens fundamental questions on nucleic acid catalysis. This chapter summarizes different strategies that are commonly used for functional characterization of enzymatic nucleic acids. Some strategies are specifically useful for enzymatic nucleic acids whereas the others are common for any type of functional nucleic acids, including aptamers and riboswitches.

2 Perturbation Assays

In many studies on functional characterization of enzymatic nucleic acids, first, the active fold of nucleic acid enzyme–substrate gets perturbed. Second, the effect of the perturbation is analyzed with a variety of biophysical and biochemical means. Perturbation of the reaction conditions such as pH, temperature, and presence or absence of specific cofactors and metal ions is frequently used for functional characterizations. The role of metal ions in catalytic activity of deoxyribozymes has been the subject of many studies [28]. Ribozymes and deoxyribozymes are known to be dependent on bivalent metal ions such as Mg^{2+}, Mn^{2+}, Hg^{2+}, and Pb^{2+} for proper folding and catalysis, and monovalent ions are also present to screen the negative charges of phosphodiester backbone. However, several recent reports emphasize the activity of some deoxyribozymes solely in the presence of monovalent metal ions such as Ag^+ [29] or Na^+ [30]. The number of trivalent-lanthanide ion-dependent deoxyribozymes is also increasing [31–34]. Besides investigation of metal ion dependence, there are reports on the addition of cationic polymers [35], organic solvents [36], or micromolar concentrations of specific metal ions. Such conditions could accelerate the enzymatic activity of deoxyribozymes up to 10,000-fold [37].

In addition, identification of critical moieties in catalysis is an important question for many studies. To this end, the molecular setup of the reaction is perturbed.

In the case of enzymatic nucleic acids, moieties such as nucleotides, nucleobases, individual functional groups, bridging or non-bridging oxygens of phosphodiester bonds, and sugar moieties can be targets of alteration. The mutation or modification may be present in the enzymatic part of the complex [38], in the substrate [19], or in both [3].

2.1 Perturbation of the Molecular Setup

2.1.1 Deletion

Deletion analysis is the first line to investigate the importance of a nucleotide [39], a small patch of nucleotides [40], or even larger domains of functional nucleic acids [41, 42]. Deletion analysis of individual nucleotides in the catalytic core of the nucleic acid enzyme would directly reveal the importance of each nucleotide as it was reported for the deoxyribozyme 10–23 [39]. The RNA-cleaving deoxyribozyme 10–23 has a small catalytic core with only 15 nucleotides. Systematic deletion analysis showed that two nucleotides of this small catalytic core (C7/T8) are not necessary for catalysis and can be omitted, albeit imposing small effects on the multi-turnover catalytic rate of the enzyme [39].

Stepwise deletion of small patches of nucleotides, so called "Deletion walk," was designed by Li and coworkers for pH5DZ1 deoxyribozyme. The deoxyribozyme pH5DZ1 is an RNA-ligating deoxyribozyme with optimal pH at 5.0 [40]. In their report, Li and colleagues analyzed the effect of the deletion of runs of 3 consecutive nucleotides in several steps, resulting in 20 mutants in a row that covered 60 nucleotides of the catalytic core. Five of the trinucleotide deletion mutants were active in a row, suggesting that the 15 nucleotides omitted in these 5 mutants are not involved in catalysis. Their hypothesis was supported with an active mutant that lacked all 15 nucleotides.

The enzyme pH5DZ1 is not the only nucleic acid enzyme that could be effectively minimized. Deletion analysis of several natural and artificial enzymatic nucleic acids led to minimized variants with almost equal activities [43–45]. The minimized form is especially useful for designing crystallization constructs [46] and spectroscopic studies [44].

Deletion of larger domains of catalytic core is common for natural ribozymes and riboswitches which are intrinsically too large to be a subject of single point deletion or deletion walk analysis. For large functional nucleic acids, domains of enzymatic nucleic acids are deleted and are investigated as in the study of group II intron by Pyle and coworkers [41] and of thiamine pyrophosphate (TPP) riboswitch by Li and Breaker [42].

The detrimental effect of a deletion might be deduced in two ways; either the functional groups of the nucleobase were necessary for catalysis or the nucleotide was only serving as a linker between the two flanking parts of the catalytic core. To unravel such ambiguity, mutation analysis would help. If a point mutant at the site of the deletion is also inactive, the first hypothesis seems to be true, otherwise, the nucleotide(s) only serves as a linker and its identity is not important for catalysis. For instance, within the catalytic core of 6BX22, an RNA-lariat forming deoxyribozyme, one deletion mutant (6BX22/min^{-1}) was shown to be inactive. However, addition of one thymidine (as a point mutation) returned the activity of the minimized enzyme (6BX22/min) [44]. The authors deduced that the identity of the deleted nucleotides was not crucial for catalysis; however, one nucleotide must be present at the deletion point to coordinate the relative positions of the two flanking domains.

In addition to standard deletion mutants that lack one or several nucleotides, deletion of functional groups of a nucleotide provides further mechanistic detail at the atomic level. Reports are present for deletion of functional groups of a nucleobase [38], the whole nucleobase [47], or the whole nucleoside [48]. In addition, deletion analysis can be performed in a combinatorial manner, which is explained in the next sections.

2.1.2 Mutations

Analysis of single-nucleotide point mutants is one of the first approaches for characterization of enzymatic nucleic acids. A detrimental mutation elucidates a critical role for the parent nucleotide in catalysis or folding [49]. In addition, in cases where a stem-loop structures or a base pair between two nucleotides is predictable, compensatory mutation to a different base pair is analyzed as in studies of 8–17 and 9DB1 deoxyribozymes [45].

The examples above emphasize the mutation analysis of the catalytic core. In addition, mutation analysis is helpful to study substrate selectivity of enzymatic nucleic acids. For example, an important aspect of RNA ligating and cleaving deoxyribozymes is sequence specificity. The identity of the dinucleotide junction at the cleavage site in cleaving deoxyribozymes and the identity of the ligation point in ligating deoxyribozymes affect efficiency of the reaction. For 2′,5′-branched RNA forming deoxyribozymes, the identity of the adjacent nucleotides flanking the branch point is shown to be important [19]. Various studies have been performed with single point mutants or double mutants at dinucleotide junctions [50] or flanking nucleotides [19] to clarify the specificity of such enzymes.

From an application point of view, some require sequence specificity of the enzyme whereas others prefer generality. For instance, F-8 deoxyribozyme was selected to site-specifically cleave a DNA substrate at thymidine-thymidine junctions [51]. On the other hand, some RNA-cleaving deoxyribozymes, "RNase DNAzymes" [52], are used as a tool for cleavage, simply as scissors. In this respect, the more general deoxyribozyme would be beneficial. A general RNase DNAzyme might be designed for any position with complementary binding arms. Although no RNase DNAzymes capable of cleavage of all junctions has been reported yet, scientists have pursued a collection of deoxyribozymes with which cleavage at all possible 16 dinucleotide junctions are possible [50, 53–55].

As mentioned above, analyzing single point or double mutants is a traditional approach to clarify the conserved sequence of functional molecules. Analyzing all possible mutants is a tedious and time-consuming approach. Recently, bioinformatics, next generation sequencing, and combinatorial methods accompanied mutation analysis to unravel the conserved sequences more efficiently and rapidly. In aptamer SELEX technology, early analysis of the selection pool with next generation sequencing methods is currently a promising approach to discover conserved sequences and binding motives [56]. The same approach is applicable for in vitro selected enzymatic nucleic acids. Reselection of a synthetic nucleic acid enzyme with a biased pool and subsequent analysis of the conserved sequences is another promising strategy [40]. In some cases, the selected clones fall into groups with clear sequences or structural homology. Sequence analysis of the selected libraries, for example, using vector NT1, directly gives an insight into the conserved sequence for catalysis [30].

2.1.3 Nucleobase Modification

Functional groups of nucleobases serve their characteristic role in catalysis and folding by coordinating metal ions [57] or forming nucleobase–nucleobase or phosphoryl oxygens–nucleobase hydrogen bonds [58]. Standard Watson–Crick base pairing, Hoogsteen interactions, and formation of G-quartets and wobble pairs are common examples of such interactions. To boost the interaction space, several newly emerging functional nucleic acids have been selected with modified nucleobases [59]. Recently, in vitro selections using modified nucleobases, for example, artificially expanded genetic information system (AEGIS) [60] or orthogonal base pairs with hydrophobic interactions [61], got more attention in expanding the functional scope and stability of enzymatic nucleic acids and aptamers. In addition, some reports focus more on backbone modification [3, 59, 62] which is explained in the next section.

   To elucidate the role of nucleobases, some studies delete the whole nucleobase and some perform "atomic mutagenesis" [57]. When the removal of the nucleobase was detrimental to catalysis, the effect of the abasic site might be rescued with exogenous nucleobases that compensate for the missing nucleobase [63]. Insertion, deletion, or replacement of functional groups has been investigated in atomic mutagenesis. Atomic mutagenesis is an informative approach in three ways; to elucidate the role of special functional group [33], to investigate the impact of the syn-anti conformation of nucleobases [58, 64], or to facilitate subsequent spectroscopic studies or crosslinking experiments [47, 65].

Elucidating the Role of Functional Groups in Ribozymes

Different types of modifications of nucleobases have been used to investigate the role of specific functional groups in ribozymes. The list of modified nucleotides that have been used for analysis of ribozymes is very extensive. A few examples are listed in Fig. 1. Modified guanosines such as 7-deazaguanosine (c^7G), 7-methylguanosine (m^7G), inosine (I), 7-deazainosine (c^7I), and 8-aza-7-

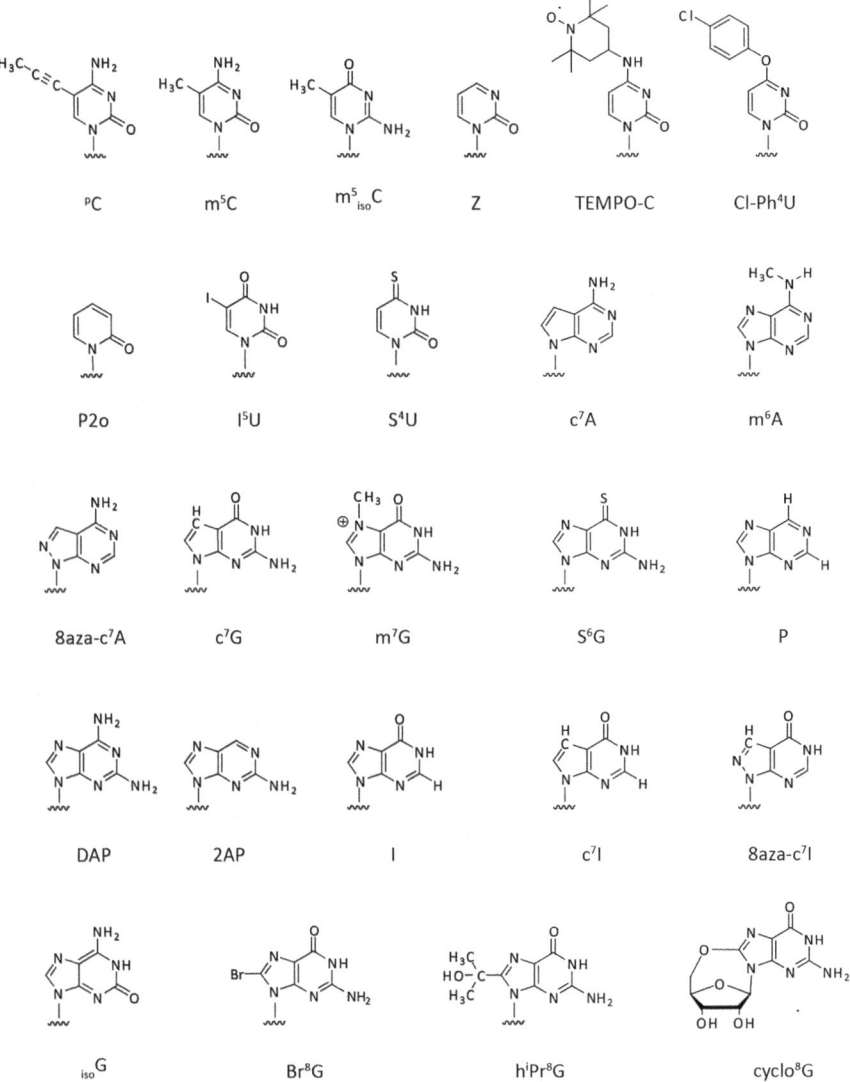

Fig. 1 Structures of the common modified nucleobases to study functional nucleic acids: 5-propyncytidine (^{P}C), 5-methylcytidine (m^{5}C), 5-methylisocytidine (m$^5_{iso}$C), zebularine (Z), 2,2,6,6-tetramethylpiperidin-1-oxyl (TEMPO)-cytidine (TEMPO-C), convertible (4-chlorophenyl)uridine (Cl-Ph^{4}U), pyridine-2-one (P2o), 5-iodouracil (I^{5}U), 4-thiourdine (s^{4}U), 7-deazaadenine (c^{7}A), N6-methyladenosine(m^{6}A), 8-aza-7-deaza-adenosine and its variants (8-aza-c^{7}A), 7-deazaguanosine (c^{7}G), 7-methylguanosine (m^{7}G), 6-thioguanosine(s^{6}G), purine (P), 2,6-diaminopurine (DAP), 2-aminopurine (2AP), inosine (I), 7-deazainosine (c^{7}I), 8-aza-7-deazainosine (8-aza-c^{7}I), isoguanosine ($_{iso}$G), 8-bromoguanosine (Br^{8}G), 8-(alpha-hydroxyisopropyl) guanosine (h^{i}Pr^{8}G), 8-5'-O-cycloguanosine (cyclo^{8}G)

deazainosine (8-aza-c^{7}I) have been utilized to study group I intron ribozyme from 1984 to date [57]. Modified forms of cytosine nucleotides have also been shown to be helpful in the study of ribozymes. In class I ligase ribozyme, the cytosine at the active site was replaced with different natural nucleobases and modified cytidines such as 5-methylcytidine(m^{5}C), zebularine (Z), pyridine-2-one (P2o), and modified forms of other nucleobases such as 2,6-diaminopurine (2AA), isoguanosine ($_{iso}$G), and 4-thiourdine (s^{4}U) for atomic interrogation on the cytosine at the active site [63].

Elucidating the Role of Functional Groups in Deoxyribozymes

The modification of nucleobases was also a subject of study for deoxyribozymes. For example, modified nucleobases, especially guanine variants, have been useful in clarifying the mechanism and structure of several deoxyribozymes such as UV1C, Ce13d and RNA cleaving deoxyribozymes. Inosine and c^{7}G modifications in addition to 7-deazaadenosine (c^{7}A), 2,6-diaminopurine(DAP) and 2-aminopurine (2AP), purine, N6-methyladenosine (m^{6}A), inosine (I), 6-thioguanosine (s^{6}G), 8-aza-7-deaza-adenosine (8-aza-c^{7}A) and its variants [66], 5-methylcytidine (m^5 C), and 5-propyncytidine (^{P}C) were used to study 10–23 [67], 8–17 [49], and Ca^{2+}-dependent RNA cleaving deoxyribozyme [68]. In a case study, guanosine to inosine mutation was used for UV1C deoxyribozyme, a deoxyribozyme with photolyase activity. In this study inosine modification was used to elucidate the electron-donating guanosine(s) for the photolyase activity [43]. In another case to investigate presence of a G-quadruplex in the structure of the lanthanide-dependent deoxyribozyme Ce13d, several guanosines were substituted with inosine (I), which lacks the amino group at position 2 of the guanine and cannot support formation of a G-quartet. The impact of the I replacement was studied kinetically [33].

Conformational Studies

Some modifications help in the investigation of the nucleobase conformation in the active fold. In a study of leadzyme, 8-bromoguanosine (Br^{8}G) was site-specifically introduced at the position of three guanosines. Br^{8}G prefers the syn conformation. With kinetic analysis, Yajima et al. revealed positions at which the presence of the syn G conformer (Br^{8}G) causes acceleration or partial inhibition of the enzyme. In another case study on group I intron, 8-5′-O-cycloguanosine (cyclo^{8}G) and 8-(alpha-hydroxyisopropyl) guanosine (h^{i}Pr^{8}G) have been site-specifically introduced at the cleavage site. The former is locked in the anti conformation and allowed cleavage. However, the latter, which is locked in the syn conformation, was detrimental to cleavage [58].

Facilitating Subsequent Studies

In some studies, modifications are aimed at facilitating subsequent spectroscopic studies or crosslinking experiments. Fluorescent characteristics of the guanine analogue 2-aminopurine (2AP) has been utilized for studying global folding kinetics of ribozymes such as HDV ribozyme [69] and group I intron [70] using steady-state and time-resolved fluorescence spectroscopic methods. In a case study, 2-aminopurine, 7-deazaguanine, and 7-deazaadenine were replaced with natural nucleobases. The latter two were used as a quencher for 2-aminopurines. Combination of such modifications enabled the study of ultrafast quenching dynamics with fluorimetry and anisotropy for leadzyme [65]. In addition to fluorescent nucleobases, other spectroscopically active moieties can be introduced to the sequence to facilitate studies such as electron paramagnetic resonance. The proof of principle has been shown with installation of 2,2,6,6-tetramethylpiperidin-1-oxyl (TEMPO) on a cytosine within two riboswitches [71].

To achieve structural data on active fold, some studies perform crosslinking reactions. Thio- and halogen-substituted nucleobases such as 6-thioguanosine [72] or 5-iodouracil [43] are introduced to the sequence of the functional nucleic acids in a site-specific manner. Subsequently, UV irradiation generates contact cross links with vicinal positions. Crosslinked complexes are purified by denaturing PAGE and analysis is performed by primer extension or piperidine cleavage [72].

In principle, any modification that alters coordination or hydrogen bonding of the functional groups of the nucleobases and can be incorporated site-specifically would be a useful modification in this respect. Some examples of such modifications have been represented above. Site-specific modifications are generally incorporated during solid-phase synthesis with phosphoramidites as building blocks. Commercial availability and the possibility of synthesis are the two criteria for selection of a modification. In some cases, a convertible nucleotide is incorporated site-specifically. The convertible nucleotide subsequently undergoes a substitution reaction to bear the desired modification [71]. Post-synthetic modifications for functional RNAs including ribozymes are also possible, using 10DM24 deoxyribozyme and modified GTPs [19]. Utilization of DNA and RNA polymerases that accept modifications or applying chemical substances that modify nucleobases within an oligo leads to random incorporation of a modification and is applicable to combinatorial analysis of the modification which is explained in the next sections.

2.1.4 Backbone Modification

In many cases of enzymatic nucleic acids, especially natural nucleolytic RNA cleaving ribozymes, the phosphodiester-sugar backbone plays a substantial role in catalysis and folding. The impact of oxygen atoms of phosphodiester bonds, sugar hydroxyls, sugar conformation and puckering, and spatial positioning of nucleobases can be investigated with backbone modifications. Deletion of some carbon atoms in the ribose ring, removal of a hydroxyl group, incorporation of locked nucleic acids (LNA) or iso-nucleic acids (isoNA, nucleotides with N-glycosidic bonds at different positions of the sugar) [62], incorporation of phosphorothioate moieties, and subsequent metal ion rescue experiments are just a few examples of this approach. In addition, backbone modifications are utilized to expand the applicability of nucleic acid catalysts. For instance, nuclease resistance and attenuation of the immune system of the host organism are two requirements for therapeutic application of nucleic acid enzymes. These two goals are achieved via modification of the phosphodiester-sugar backbone in many studies [59].

Locked nucleic acids contain a ribose moiety with a linkage between C2′ and C4′. Such a carbon–carbon bond locks the ribose in the C3′-endo pucker conformation – hence the name "locked." LNAs are mainly applied to increase nuclease resistance of oligonucleotides in vivo [73]. Site-specific presence of LNA in a DNA or RNA backbone restricts the ribose sugar conformation at the introduction site. When the ribose pucker plays a role in catalysis, this restriction may either accelerate or inhibit the enzymatic activity. The case was observed in leadzyme and was attributed to the C3′-endo conformation of specific guanosines being involved in the formation of a transition state [74]. One should note that LNA lack the 2′-OH and the effect must be dissected with control experiments using 2′-O-methyl or 2′-fluoro nucleotides [74].

Phosphorothioate modification involves substituting a non-bridging oxygen atom by a sulfur atom. Oligonucleotides with site-specific phosphorothioate substitutions are synthesized by solid phase. Subsequently, the S_P and R_P diastereomers can be separated by C18 reverse-phase HPLC [75]. In many cases, presence of the thio group diminishes the catalytic activity [33, 76]. For catalytic nucleic acids, the loss of activity of either of the S_P or R_P diastereomers may be rescued in the presence of a thiophilic metal ion such as Hg^{2+}, Pb^{2+}, and Cd^{2+} [63,75], although the metal ion rescue is not granted for all catalytic nucleic acids [77]. The thiophilic metal ion rescue of phosphorothioate backbone was used as an advantage for development of a sensor for heavy metal ions with deoxyribozymes [78].

Accepting modifications that make deoxyribozymes resistant to endo- and exonucleases and inhibits immune response of target organisms allows in vivo and therapeutic application of deoxyribozymes [59]. This type of modifications include 2′-O-methyl, 2′-fluoro, and 2′-amino modifications [73, 79, 80], LNA [81], N3′-P5′ phosphoramidite [82], phosphorothioates [83], synthetic genetic polymers such as cyclohexene nucleic acids (CeNA), hexitol nucleic acids (HNA), arabino nucleic acids (ANA) and 2′-fluoroarabino nucleic acids (FANA) [3], threose nucleic acids (TNA) [84], peptide nucleic acids (PNA), and 3′-3′ inverted nucleoside.

2.2   Biochemical and Biophysical Assays

Perturbing molecular setup or reaction condition is usually followed by investigation with a combination of biochemical and biophysical methods. For enzymatic nucleic acids, kinetic analysis is the first line for such investigations [28]. In many cases, the presence of specific metal ions [75] or exogenous nucleosides [47] rescued the activity of the perturbed system. In these cases, kinetic studies and calculation of the relative rate constants revealed mechanistic details of the system.

Enzymatic nucleic acids mostly catalyze single turnover reactions. However, some self-cleaving deoxyribozymes were shown to catalyze multi-turnover reactions under specific conditions, for example, having short binding arms [85]. Most studies assume single-step catalysis and calculate and compare observed rate constants (k_{obs}). In contrast, some enzymatic nucleic acids are known to follow double-step mechanisms [16]. In a two-step mechanism, first, an intermediate product is produced and accumulates and then the intermediate gets used in the second step of the reaction. The rate constant for each step is measureable by kinetic analysis [65].

Most substrates and products of enzymatic nucleic acids are nucleic acids. Thus, the product formation in the course of reaction can be visualized by gel shift in denaturing urea-PAGE or ion exchange HPLC with a standard UV detector. In the former, one substrate and the product are mostly radioactively labeled. Autoradiography or scanning a PhosphorImager screen with a laser reader is used to visualize the bands. In addition to radioactive set-ups, fluorescent labels are also possible. After scanning, the band intensities are analyzed and are integrated by software such as ImageQuant or ImageJ. In the latter, the chromatogram outcome of the HPLC is analyzed and band intensities of special peaks are integrated for analysis.

Besides kinetic studies, probing assays are useful to compare a standard condition with a perturbed condition. Many probing assays are available; examples include DMS [86], Fenton [87], and DNase I [88] assays. Selective 2'-hydroxyl acylation analyzed by primer extension (SHAPE) [89], in-line probing [75] terbium (III) footprinting [69], and crosslinking using iodine [90] are the usual additional probing assays.

Spectroscopic methods such as UV melting [65], circular dichroism (CD) [44, 69], fluorimetry, anisotropy [65], Förster resonance energy transfer (FRET) [69], terbium luminescent studies [91], and nuclear magnetic resonance (NMR) are additional approaches that demonstrate the global conformational changes during catalysis. Each method has its advantages and limitations. Among the spectroscopic methods, NMR is only useful for small enzymatic nucleic acids [92] or small parts of a larger one [93]. Fluorimetric analysis requires the presence of fluorescent moieties that have been site-specifically incorporated within the sequence of the enzymatic nucleic acid or its substrate [65].

Biochemical and biophysical studies are best validated with X-ray crystallography. Crystal structures provide a strong basis to explain biochemical data and understand the functionality of enzymatic nucleic acids. Today, high-resolution crystal structures of almost all the natural small endonucleolytic ribozymes [8, 11, 94–96] and many synthetic ribozymes [15, 97–100] are available. In addition, the first crystal structure of an active fold of a deoxyribozyme was recently reported by Höbartner and colleagues on the post-catalytic state of deoxyribozyme 9DB1

[46]. After 1997, a crystal structure of a catalytically inactive fold of an RNA-cleaving deoxyribozyme was published [101], the first report that showed consistency with biochemical investigations. 9DB1 catalyzes nucleophilic attack of a 3'-OH of an RNA substrate to a 5'-triphosphate of another RNA substrate and forms a linear ligated RNA product [46].

X-Ray crystallography, along with several other biochemical, biophysical, and bioinformatic methods and known propensities of nucleobases for complex formation, makes a powerful basis for proposing the catalytic mechanism of enzymatic nucleic acids [102] and their rational mutagenesis [46]. The availability of the crystal structure of a system directs intelligent perturbations [46] and deduction of the results [103]. The availability of crystallography data grants atomic understanding of the catalytic mechanisms of similar systems [11]. However, crystal structures are not always available, especially for deoxyribozymes, and in many cases perturbation assays are the only means to understand the catalytic mechanism of enzymatic nucleic acids.

3   Combinatorial Approaches

Analyzing all possible single point mutants and modified forms of functional nucleic acids is a laborious and time-consuming task. In a study about the small deoxyribozyme 8–17, all three possible mutations for 9 positions of the catalytic core were kinetically studied under 4 different conditions, meaning synthesis of 27 different deoxyribozymes and performing almost 120 kinetic experiments [49]. For longer nucleic acids this type of study would be more backbreaking. Combinatorial approaches for simultaneous and yet individual analyses of perturbations are worthy solutions to this problem.

There are several combinatorial mutation/modification reports in the literature. Nucleotide analogue interference mapping (NAIM) [104] and enzymatic combinatorial nucleoside deletion scanning (enzymatic NDS) [105] are specifically designed for functional RNAs including ribozymes. Combinatorial mutation interference analysis (CoMA) [45], nucleotide analogue interference mapping for DNA (dNAIM) [38], combinatorial nucleoside deletion scanning (NDS) [48], and dimethyl sulfate interference mapping (DMSi) [44, 50] are applied to functional DNAs, including deoxyribozymes. All methods share a common four-step concept: (1) distribution of tagged perturbations within the sequence, (2) separation of active species, (3) tag-specific reactions, and (4) sequencing gel and analysis.

In step one, the perturbation of interest, that is, mutations or modifications, are introduced within the sequence of functional nucleic acids. This step forms a "library" or "pool" of nucleic acids with the desired alteration. The introduction of the perturbations might be simultaneous with synthesis or post-synthesis. It might also be enzymatic or chemical. Within the methods mentioned here, NAIM benefits from enzymatic incorporation of modified tagged nucleosides into RNA, whereas CoMA, NDS, dNAM, and enzymatic NDS incorporate the modification

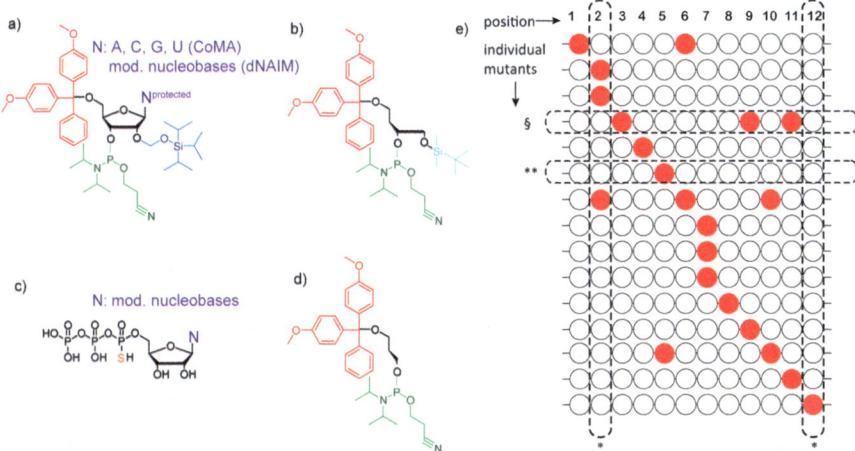

Fig. 2 Building blocks for combinatorial methods. (**a**) Standard 3′-β-cyanoethyl phosphoramidites for CoMA and dNAIM, *dark blue*: TOM protecting group (2-*O*-triisopropylsilyloxymethyl) on 2′ position, *red*: dimethoxytrityl protecting group (DMT) on 5′ position, *green*: β-cyanoethyl phosphoramidites. (**b**) Butantriol-derived spacer for NDS, *light blue*: *tert*-butyldimethylsilyl (*t*BDMS) protecting group on 2′ position. (**c**) α-Thionucleotides for NAIM. (**d**) Spacer building blocks for enzymatic NDS. (**e**) Simplified and schematic depiction of a perturbation library for a 12-mer oligonucleotide. *Each position have a subpopulation of mutants. Here subpopulations for position 2 and 12 are marked with an *asterisk*. **Most mutants only contain one mutation. The individual mutants marked with *two asterisks* contain a mutation at position 5. § Minority of mutants bear more than one mutation or no mutation. Here, the individual molecule with § mark contains three mutations at positions 3, 9, and 11

via solid phase synthesis. The method DMSi forms the modifications post-synthetically and chemically. Figure 2a–d illustrates the building blocks used to synthesize libraries for CoMA, dNAIM, NDS, NAIM, and enzymatic NDS.

In general, the NAIM method is designed for functional RNAs and, since its introduction by Strobel and Shetty in 1997 on group I intron [106], has been used for many studies [107, 108]. The NAIM protocol has been well-explained recently by Nilsen [104]. In the first step of NAIM, modified α-thionucleotides are incorporated into the RNA of interest. Incorporation of α-thionucleotides are enzymatic – applying T7 RNA polymerase. Upon nucleotide incorporation, pyrophosphates are released and α-thiophosphate remains as the linking thiophosphate backbone of RNA. The α-thiophosphate group is the tag with which the presence of individual modified species are probed during the third step. Following the introduction of the NAIM method, many studies have reported base modifications acceptable for T7 RNA polymerase and can be used for NAIM analysis. Examples include 2,6-diaminopurine (DAP), 2-aminopurine (2AP), purine (P), 7-deazaadenosine (c^7A), and N6-methyladenosine (m^6A) [104].

In contrast to NAIM, incorporation of perturbations for CoMA, NDS, enzymatic NDS, and dNAIM is via chemical solid phase synthesis. Phosphoramidites bearing the perturbation of interest are either purchased from commercial providers or synthesized by laboratories. CoMA requires synthesis of four libraries, each with one standard nucleotides chemically tagged with a 2'-hydroxyl, namely rA, rG, rC, and rU. NDS requires the synthesis of libraries with statistically dispersed non-nucleosidic spacer C4 units, that is, 1,2,4-butantriol derivative. The C4 spacers preserves an OH group, resembling the 2'-OH of the ribose sugar as a chemical tag. Enzymatic NDS is dependent on synthesis of libraries with non-nucleosidic C3 spacer units that lack the hydroxyl group of 2' position of the ribose sugar. In the dNAIM method, libraries of modified nucleobases chemically tagged with 2'-OH are synthesized. Finally, the method DMSi applies dimethyl sulfate (DMS) post-synthesis to methylate N7 positions of guanosines within the DNA of interest. DNA is synthesized by solid phase synthesis and is then subjected to optimized concentrations of DMS.

The method of incorporation of perturbation must be optimized in such a way that each individual position of the nucleic acid of interest has a sub-population of modified forms. Simultaneously, most sub-populations must only contain one modification, neither less nor more. The distribution of DNA molecules containing defined numbers of mutations can be calculated by a binomial distribution [45]. Figure 2e depicts a simplified example of dispersed perturbations of a library in the combinatorial methods.

To achieve this goal, NAIM uses a mixture of standard NTPs with modified α-thionucleotides. Four libraries of ribo A, G, C, and U are synthesized by RNA polymerase. Each library bears modified α-thionucleotides accordingly. As an example, the A-modified library is synthesized with standard GTP, CTP, and UTP and a mixture of ATP and α-thio-modified ATP. A control library is also synthesized with standard GTP, CTP, UTP, and α-thio-non-modified ATP. According to the template DNA, the RNA polymerase incorporates either of ATP or α-thio-modified ATP in front of thymidines. The ratio of incorporation of ATP vs α-thio-modified ATP is controlled by percentage of α-thio-modified ATP. Additionally, because the rate of incorporation of ATP and α-thio-modified ATP by RNA polymerase is different, the preference of the enzyme for ATP incorporation over α-thio-modified ATP must be taken into account. Thus, to achieve a desired library with evenly distributed modifications, the effective percentage of ATP vs α-thio-modified ATP must be noted.

In the cases of CoMA, dNAIM, NDS, and enzymatic NDS, the libraries are synthesized on solid phase with phosphoramidite chemistry. In these cases, the ratio of standard parent phosphoramidites vs modified phosphoramidite controls the distribution of sub-populations. As described above for NAIM, the incorporation of standard phosphoramidites and modified phosphoramidites are not equally efficient. This means that the actual molar ratio of phosphoramidites does not necessarily reflect the incorporation ratio. Comprehensive studies have been done by Höbartner and colleagues to achieve calibration curves for phosphoramidite incorporations and their modified forms [45, 48, 105].

For DMSi, the ratio of methylated forms of DNA at individual positions is controlled by concentration of the chemical reagent DMS and reaction times. To stop the reaction at a desired time point, a quenching solution, containing β-mercaptoethanol and NaCl is used. Quenching is followed by ethanol precipitation and removal of excess DMS.

Step two of the combinatorial methods is separation of active nucleic acids. Active sub-populations are separated from inactive or not yet active sub-populations. During this step, the perturbed libraries become involved in the catalytic reaction. The reaction setup is designed in such a way that the active species alter themselves during catalysis. This self-alteration is used as a means for separation. For instance, for deoxyribozymes that cleave an oligonucleotide, the substrate is covalently ligated to the library, forming a uni-molecular enzyme–substrate complex (Fig. 3b). The covalent linkage between the enzyme and substrate is formed with T4 RNA ligase (5′ end) or T4 DNA ligase (3′ end, splint mediated). The uni-molecular enzyme–substrate complex undergoes cleavage reaction. The cleavage of the substrate reduces the molecular weight of the enzyme–substrate complex (Fig. 3c). Thus, the active species alter their own molecular weight upon catalysis and can be separated from the non-reacted species by a denaturing polyacrylamide gel electrophoresis (Fig. 3d). The active fraction is cut from gel and the mixture of active mutants is purified from gel by a crush and soak method followed by ethanol precipitation. Obviously, for many ribozymes with self-cleavage activity, the ligation of the substrate (with T4 RNA/DNA ligase) is not required. For ligating deoxyribozymes, one substrate is ligated covalently to the library and the reactions are formed in bi-molecular formats. The separation step may also be based on affinity binding. The cleaved sequence may bear a binding motif such as an aptamer that is missed after cleavage by active species.

Separation results in two fractions; the active fraction and the inactive plus the not yet reacted fraction. The active fraction contains a mixture of species that tolerate both chemical tags and mutation or modifications. However, not all species that tolerate the introduced chemical tag and perturbation end in the active fraction. Based on the time given for the reaction, some slow-catalyzing species may not get time to react. In addition, some sub-population of perturbed enzymes may get misfolded transiently and cannot react within the time scale of the reaction. Therefore, although one fraction is named as the active fraction, the remaining fraction should be mentioned as a mixture of inactive species plus species that could not yet catalyze. Thus, in the third step, the active fraction is analyzed and compared to the unseparated library.

The separation step is usually performed in cold fashion in 1–2 nanomoles of reactants. To get a well-resolved sequencing gel at the last step, a fraction of 50–100 picomoles of the enzyme–substrate complex are trace labeled with radioactive phosphate prior to the third step. The label is installed either at the 5′-end with [γ-^{32}P] ATP and T4 polynucleotide kinase [109] or at the 3′-end with [5′-^{32}P] cytidine 3′,5′-bisphosphate, that is, [^{32}P]pCp [110].

Step three of the combinatorial methods implies the tags to visualize at which positions the modifications and mutations are allowed. The active fraction (experiment) and unseparated libraries (control) are subjected to tag-specific reactions. Individual species within the mixture of active fractions are cleaved (Fig. 3e). The oligonucleotides have been radiolabeled at one end. Therefore the cleavage at

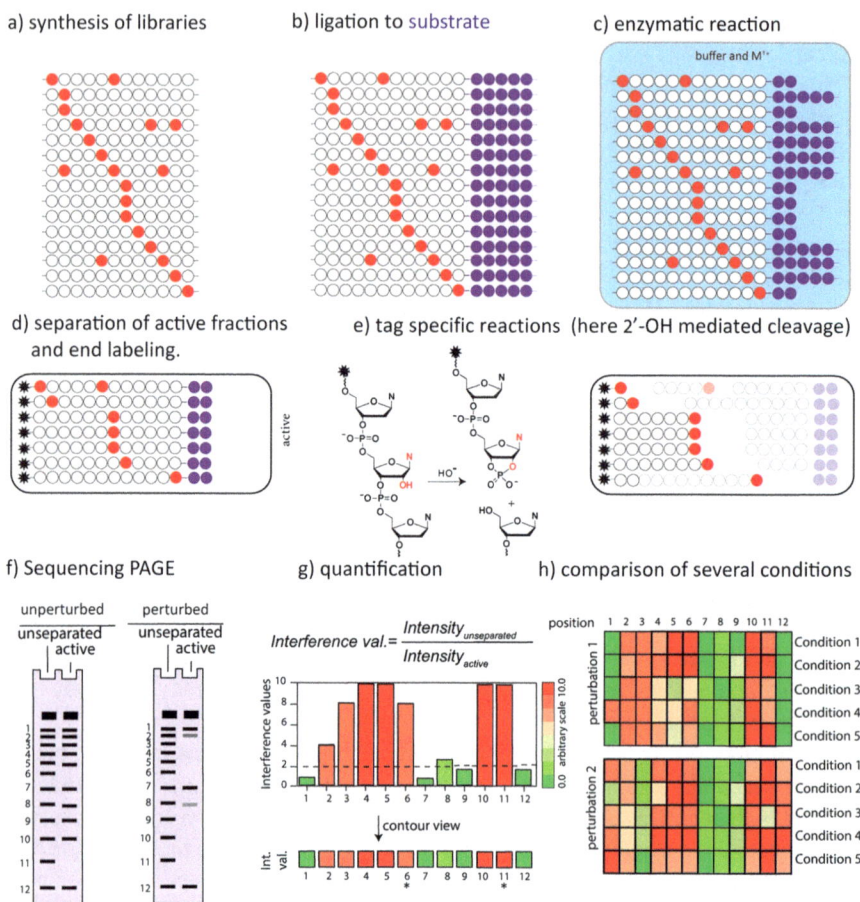

Fig. 3 General workflow for combinatorial perturbation analyses for an oligonucleotide-cleaving nucleic acid. (**a**) Synthesis of perturbed libraries by chemical or enzymatic means. The tagged perturbations (*red circles*) are statistically dispersed within the functional segments. (**b**) Ligation of the substrate to the library, typically with T4 RNA or DNA ligase. (**c**) Addition of buffer and metal ions to start cleavage reaction of the enzymatic nucleic acid. (**d**) Separation of active fraction and radiolabeling at one end. (**e**) Tag-specific reaction, for example, for CoMA, dNAIM, and NDS – an inline attack of the 2'-OH chemical tag. (**f**) Sequencing urea-PAGE. (**g**) Quantification and calculation of the interference values. *Asterisks* indicate positions at which the respective band is absent in the active fraction of the unperturbed library and mean detrimental effect of the chemical tag. (**h**) Several perturbations and different conditions may be compared with contour graphs. Here, positions 7–9 accept both perturbation in all 5 conditions, suggesting that the region may serve as a linker between two flanking parts

perturbation sites produces a ladder from present oligos. The produced ladder is resolved on a sequencing gel in step four.

As mentioned above, the chemical tag for NAIM is an α-thiophosphate group. The thiophosphate group is sensitive for cleavage in presence of iodoethanol. Thus,

Table 1 Chemical tag and tag-specific reactions of combinatorial methods to examine functional RNA and DNAs

Method	Subject of study	Perturbation tag	Tag-specific reaction
NAIM	RNA	α-Thiophosphate	Iodoethanol cleavage
Enzymatic NDS	RNA	Missing 2'-OH	Aborted primer extension
CoMA	DNA	2'-OH	In-line attack by 2'-OH
dNAIM	DNA	2'-OH	In-line attack by 2'-OH
NDS	DNA	2'-OH	In-line attack by 2'-OH
DMSi	DNA	Methyl group	Piperidine cleavage

sub-molar concentrations of iodine in ethanol are used for site-specific cleavage at modification sites. CoMA, dNAIM, and NDS are designed to study functional DNAs. With a background of deoxyriboses, the 2'-OH chemical tag allows application of alkaline hydrolysis for site-specific cleavage at perturbation sites. The enzymatic NDS method is designed to analyze functional RNAs. Missing the 2'-OH is the chemical tag for this method. The third step of enzymatic NDS applies primer extension with M-MuLV and SSIII reverse transcriptases. These reverse transcriptase enzymes could not pass the deoxy C3 spacer within the RNA of the template and are aborted at the perturbations sites. Finally, the methylation sites for DMSi are cleaved with the addition of piperidine (10% in water) and incubation at 90 °C (Table 1).

Step four uses denaturing PAGE to resolve the cleaved products of NAIM, CoMA, dNAIM, NDS enzymatic DNA and DMSi, and aborted primer extension products of enzymatic NDS. Species in the active fraction are oligonucleotides that tolerate perturbation of interest and form a ladder in the sequencing gel. In contrast, positions at which the perturbation of interest is not allowed do not show the relevant band and are reflected in the sequencing gel as missing bands. The libraries are radiolabeled and the bands are visualized by autoradiography or using a PhosphorImager.

To quantify the interference effect of the perturbation, the band intensities at individual positions of the active fraction and unseparated libraries are compared. Dividing the band intensity of every nucleotide position in the unseparated library by the band intensity in the active fraction gives interference values for each position. An interference value of 1 means no effect is caused by the perturbation. Higher interference values mean that the perturbations were interfering. However, interference values between 1 and 2 are negligible. Interference values above 10 are rounded as 10.

In the combinatorial methods mentioned, the perturbations were accompanied with a tag. The tags may impose electrostatic and steric hindrance in sensitive positions. The effect of a tag on catalysis must be dissected from the effect of the perturbation of interest. Thiophosphate interference lanes in NAIM and hydroxyl lanes in dNAIM are controls for catalytic impacts of the chemical tag. For instance, when the effect of a modification on adenosine is the subject of study, beside the active lane and unselected lane for the modified adenosine, one lane is run with unmodified adenosine bearing only the thiophosphate tag. Missing a band in the unmodified adenosine lane reflects the tag effect at a specific position. At such a position NAIM is not informative. The same is valid for studying modifications with dNAIM. For CoMA, each mutant library reflects the mutation effect for three nucleobase and the tag effect for the parent nucleobase, that is, the cleaved products of the rA library reflect the mutation effect at positions that were initially C, G, and T and the tag effect at positions that were initially A. Thus, a missing band at the parent position in any of the four libraries reflects the detrimental effect of the 2'-OH at that specific position. Figure 3 illustrates the overall scheme for combinatorial perturbation methods.

In summary, the combinatorial perturbation methods give simultaneous insights for individual modifications and mutations within a single polyacrylamide gel. Knowing tolerant positions of a nucleic acid enzyme to perturbations helps rational engineering. Several combinatorial perturbation studies led to trimming of deoxyribozymes to 60–80% of the original length [44], mutation cycle analysis [48], and directed mutagenesis [46]. The combinatorial approach for perturbation analysis provides a good platform to conjugate deoxyribozymes and aptamers as novel molecular sensors. Development and design of such splitzymes [111] are more straightforward if positions that tolerate mutations are identified. In many studies on deoxyribozymes, a stretch of consecutive nucleotides were mutable. It was shown that such regions within the catalytic core of deoxyribozymes serve only as a linker and can be replaced by nucleotides that were not originally in the sequence [44]. These linker sites provide a good target for insertion of aptamers and other functional sequences. The linker sites may also tolerate modifications, for example, biotinylation, amination, and thiolation that facilitate the application of enzymatic nucleic acids in different systems such as biosensors.

4   Conclusion

For many ribozymes perturbation analyses, biophysical and biochemical studies and crystal structures confirmed the role of the 2'-hydroxyl for catalysis. In contrast, the catalytic mechanism of deoxyribozymes was unknown for many years in the absence of crystal structures. With the first crystallography report on 9DB1 deoxyribozyme, hopes to achieve further deoxyribozyme structures and a clear understanding of their mechanism are elevated.

In the absence of crystal structures, comparison of the data from several different perturbation approaches unravels insights into catalytic mechanism of nucleic acids. Deletion, mutation, and modification of nucleobases and backbones that are followed by biophysical and biochemical work such as kinetic analyses, spectroscopy, and probing assays are the core functional studies for nucleic acids. In this respect, combinatorial perturbation interference analyses reveal comprehensive information on the catalytically and structurally important nucleotides and functional groups. NAIM and enzymatic NDS investigate functional RNAs, whereas

CoMA, dNAIM, NDS, and DMSi examine functional DNAs. CoMA simultaneously reflects the effect of single point mutations. NAIM and dNAIM give atomic understanding of functional groups. NDS and enzymatic NDS provide information on the importance of nucleosides and DMSi reflects the effect of methylation at N^7 of guanosines of the catalytic core. Altogether, combinatorial perturbation approaches provide substantial data in a time-effective manner for functional nucleic acids.

Knowing the mechanistic details of enzymatic nucleic acids facilitates their engineering to be incorporated into a bio-sensing system [112, 113]. Knowledge on the acceptance of special backbone modifications in addition to studies on the metal ion requirements of in vitro selected enzymatic nucleic acids paves the way for their in vivo and therapeutic applications [80]. Functional studies also impact rational mutation and modification to expand substrate acceptability [46] or narrow substrate specificity. Kinetic studies on enzymatic nucleic acids and knowing optimal conditions for their activity improves their applications as a tool in molecular biology [37].

References

1. Sczepanski JT, Joyce GF (2014) Nature 515:440
2. Silverman SK (2016) Trends Biochem Sci 41(7):595–609
3. Taylor AI, Pinheiro VB, Smola MJ, Morgunov AS, Peak-Chew S, Cozens C, Weeks KM, Herdewijn P, Holliger P (2015) Nature 518:427
4. Martick M, Scott WG (2006) Cell 126:309
5. Rupert PB, Massey AP, Sigurdsson ST, Ferre-D'Amare AR (2002) Science 298:1421
6. Ke A, Zhou K, Ding F, Cate JH, Doudna JA (2004) Nature 429:201
7. Kennell JC, Saville BJ, Mohr S, Kuiper MT, Sabourin JR, Collins RA, Lambowitz AM (1995) Genes Dev 9:294
8. Ren A, Kosutic M, Rajashankar KR, Frener M, Santner T, Westhof E, Micura R, Patel D (2014) J Nat Commun 5:5534
9. Cochrane JC, Lipchock SV, Strobel SA (2007) Chem Biol 14:97
10. Stahley MR, Adams PL, Wang J, Strobel SA (2007) J Mol Biol 372:89
11. Peters JK, Toor N (2015) RNA Biol 12:913
12. Reiter NJ, Osterman A, Torres-Larios A, Swinger KK, Pan T, Mondragon A (2010) Nature 468:784
13. Ekland EH, Szostak JW, Bartel DP (1995) Science 269:364
14. Robertson MP, Scott WG (2007) Science 315:1549
15. Xiao H, Murakami H, Suga H, Ferre-D'Amare AR (2008) Nature 454:358
16. Qi X, Xia T (2011) Biomol Concepts 2:305
17. Silverman SK (2009) Acc Chem Res 42:1521
18. Yehl K, Joshi JP, Greene BL, Dyer RB, Nahta R, Salaita K (2012) ACS Nano 6:9150
19. Büttner L, Javadi-Zarnaghi F, Höbartner C (2014) J Am Chem Soc 136:8131
20. Elbaz J, Lioubashevski O, Wang F, Remacle F, Levine RD, Willner I (2010) Nat Nanotechnol 5:417
21. Zhang X-B, Kong R-M, Lu Y (2011) Ann Rev Anal Chem 4:105
22. Zhang W, Feng Q, Chang D, Tram K, Li Y (2016) Methods 24:30052

23. Hesser AR, Brandsen BM, Walsh SM, Wang P, Silverman SK (2016) Chem Commun 52 (59):9259–9262
24. Camden AJ, Walsh SM, Suk SH, Silverman SK (2016) Biochemistry 55:2671
25. Silverman SK (2015) Acc Chem Res 48:1369
26. Zhou C, Avins JL, Klauser PC, Brandsen BM, Lee Y, Silverman SK (2016) J Am Chem Soc 138:2106
27. Hollenstein M, Hipolito CJ, Lam CH, Perrin DM (2009) Chembiochem 10:1988
28. Bonaccio M, Credali A, Peracchi A (2004) Nucl Acids Res 32:916
29. Saran R, Liu J (2016) Anal Chem 88:4014
30. Torabi SF, Wu P, McGhee CE, Chen L, Hwang K, Zheng N, Cheng J, Lu Y (2015) Proc Natl Acad Sci USA 112:5903
31. Zhou W, Vazin M, Yu T, Ding J, Liu J (2016) Chem Eur J 22(28):9835–9840
32. Huang PJ, Vazin M, Liu J (2016) Biochemistry 55:2518
33. Vazin M, Huang PJ, Matuszek Z, Liu J (2015) Biochemistry 54:6132
34. Huang PJ, Vazin M, Liu J (2014) Anal Chem 86:9993
35. Saito K, Shimada N, Maruyama A (2016) Sci Technol Adv Mater 17:437
36. Behera AK, Schlund KJ, Mason AJ, Alila KO, Han M, Grout RL, Baum DA (2013) Biopolymers 99:382
37. Javadi-Zarnaghi F, Höbartner C J Am Chem Soc 2013
38. Wachowius F, Höbartner C (2011) J Am Chem Soc 133:14888
39. Zaborowska Z, Schubert S, Kurreck J, Erdmann VA (2005) FEBS Lett 579:554
40. Kandadai SA, Mok WW, Ali MM, Li Y (2009) Biochemistry 48:7383
41. Fedorova O, Mitros T, Pyle AM (2003) J Mol Biol 330:197
42. Li S, Breaker RR (2013) Nucl Acids Res 41:3022
43. Chinnapen DJ, Sen D (2007) J Mol Biol 365:1326
44. Javadi-Zarnaghi F, Höbartner C (2016) Chem Eur J 22:3720
45. Wachowius F, Javadi-Zarnaghi F, Höbartner C (2010) Angew Chem Int Ed 49:8504
46. Ponce-Salvatierra A, Wawrzyniak-Turek K, Steuerwald U, Höbartner C, Pena V (2016) Nature 529:231
47. Kuzmin YI, Da Costa CP, Fedor MJ (2004) J Mol Biol 340:233
48. Samanta B, Höbartner C (2013) Angew Chem Int Ed 52:2995
49. Peracchi A, Bonaccio M, Clerici M (2005) J Mol Biol 352:783
50. Lam JC, Kwan SO, Li Y (2011) Mol Biosyst 7:2139
51. Wang M, Zhang H, Zhang W, Zhao Y, Yasmeen A, Zhou L, Yu X, Tang Z (2014) Nucl Acids Res 42(14):9262–9269
52. Tram K, Kanda P, Li Y (2012) J Nucl Acids 2012:958683
53. Cruz RP, Withers JB, Li Y (2004) Chem Biol 11:57
54. Schlosser K, Gu J, Lam JC, Li Y (2008) Nucl Acids Res 36:4768
55. Schlosser K, Gu J, Sule L, Li Y (2008) Nucl Acids Res 36:1472
56. Blank M (2016) Methods Mol Biol 1380:85
57. Forconi M, Benz-Moy T, Gleitsman KR, Ruben E, Metz C, Herschlag D (2012) RNA 18:1222
58. Lin CW, Hanna M, Szostak JW (1994) Biochemistry 33:2703
59. Meek KN, Rangel AE, Heemstra JM (2016) Methods 106:29–36. ISSN 1046-2023
60. Sefah K, Yang Z, Bradley KM, Hoshika S, Jimenez E, Zhang L, Zhu G, Shanker S, Yu F, Turek D, Tan W, Benner SA (2014) Proc Natl Acad Sci U S A 111:1449
61. Li L, Degardin M, Lavergne T, Malyshev DA, Dhami K, Ordoukhanian P, Romesberg FE (2014) J Am Chem Soc 136:826
62. Yang X, Xiao Z, Zhu J, Li Z, He J, Zhang L (2016) Yang, Z. Org Biomol Chem 14:4032
63. Shechner DM, Bartel DP (2011) Nat Struct Mol Biol 18:1036
64. Yajima R, Proctor DJ, Kierzek R, Kierzek E, Bevilacqua PC (2007) Chem Biol 14:23
65. Kadakkuzha BM, Zhao L, Xia T (2009) Biochemistry 48:3807

66. He J, Zhang D, Wang Q, Wei X, Cheng M, Liu K (2011) Org Biomol Chem 9:5728
67. Zhu J, Li Z, Yang Z, He J (2015) Bioorg Med Chem 23:4256
68. Okumoto Y, Tanabe Y, Sugimoto N (2003) Biochemistry 42:2158
69. Gondert ME, Tinsley RA, Rueda D, Walter NG (2006) Biochemistry 45:7563
70. Chauhan S, Behrouzi R, Rangan P, Woodson SA (2009) J Mol Biol 386:1167
71. Büttner L, Seikowski J, Wawrzyniak K, Ochmann A, Höbartner C (2013) Bioorg Med Chem 21(20):6171–6180
72. Liu Y, Sen D (2008) J Mol Biol 381:845
73. Schubert S, Kurreck J (2004) Curr Drug Targets 5:667
74. Julien KR, Sumita M, Chen P-H, Laird-Offringa IA, Hoogstraten CG (2008) RNA 14:1632
75. Thaplyal P, Ganguly A, Golden BL, Hammes-Schiffer S, Bevilacqua PC (2013) Biochemistry 52:6499
76. Nawrot B, Widera K, Wojcik M, Rebowska B, Nowak G, Stec WJ (2007) FEBS J 274:1062
77. Huang PJ, Vazin M, Matuszek Z, Liu J (2015) Nucl Acids Res 43:461
78. Huang PJ, Liu J (2014) Anal Chem 86:5999
79. Fokina AA, Meschaninova MI, Durfort T, Venyaminova AG, Francois JC (2012) Biochemistry 51:2181
80. Fokina AA, Stetsenko DA, Francois JC (2015) Exp Opin Biol Ther 15:689
81. Doessing H, Vester B (2011) Molecules 16:4511
82. Takahashi H, Hamazaki H, Habu Y, Hayashi M, Abe T, Miyano-Kurosaki N, Takaku H (2004) FEBS Lett 560:69
83. Burnett JC, Rossi JJ (2012) Chem Biol 19:60
84. Yu H, Zhang S, Dunn MR, Chaput JC (2013) J Am Chem Soc 135:3583
85. Silverman SK (2005) Nucl Acids Res 33:6151
86. Waldsich C, Schroeder R (2008) Handbook of RNA biochemistry. Wiley-VCH Verlag GmbH, Weinheim, p 229
87. Jain SS, Tullius TD (2008) Nat Protoc 3:1092
88. Cardew AS, Fox KR (2010) Methods Mol Biol 613:153
89. Wilkinson KA, Merino EJ, Weeks KM (2006) Nat Protoc 1:1610
90. Sekhon GS, Sen D (2009) Biochemistry 48:6335
91. Kim HK, Li J, Nagraj N, Lu Y (2008) Chem Eur J 14:8696
92. Hoogstraten CG, Legault P, Pardi A (1998) J Mol Biol 284:337
93. Skilandat M, Rowinska-Zyrek M, Sigel RK (2014) J Biol Inorgan Chem 19:903
94. Eiler D, Wang J, Steitz TA (2014) Proc Natl Acad Sci U S A 111:13028
95. Liu Y, Wilson TJ, McPhee SA, Lilley DM (2014) Nat Chem Biol 10:739
96. Suslov NB, DasGupta S, Huang H, Fuller JR, Lilley DM, Rice PA, Piccirilli JA (2015) Nat Chem Biol 11:840
97. Shechner DM, Grant RA, Bagby SC, Koldobskaya Y, Piccirilli JA, Bartel DP (2009) Science 326:1271
98. Pitt JN, Ferré-D'Amaré AR (2009) J Am Chem Soc 131:3532
99. Serganov A, Keiper S, Malinina L, Tereshko V, Skripkin E, Höbartner C, Polonskaia A, Phan AT, Wombacher R, Micura R, Dauter Z, Jäschke A, Patel DJ (2005) Nat Struct Mol Biol 12:218
100. Wedekind JE, McKay DB (2003) Biochemistry 42:9554
101. Nowakowski J, Shim PJ, Prasad GS, Stout CD, Joyce GF (1999) Nat Struct Biol 6:151
102. Koo SC, Lu J, Li NS, Leung E, Das SR, Harris ME, Piccirilli JA (2015) J Am Chem Soc 137:8973
103. Chen JH, Yajima R, Chadalavada DM, Chase E, Bevilacqua PC, Golden BL (2010) Biochemistry 49:6508
104. Nilsen TW (2015) Cold Spring Harb Protoc 2015:604
105. Wawrzyniak-Turek K, Höbartner C (2014) Chem Commun 50:10937

106. Strobel SA, Shetty K (1997) Proc Natl Acad Sci U S A 94:2903
107. Waldsich C (2008) Nat Protoc 3:811
108. Jansen JA, McCarthy TJ, Soukup GA, Soukup JK (2006) Nat Struct Mol Biol 13:517
109. Rio DC (2014) Cold Spring Harb Protoc 2014:441
110. Nilsen TW (2014) Cold Spring Harb Protoc 2014:444
111. Alila KO, Baum DA (2011) Chem Commun 47:3227
112. Wang F, Elbaz J, Teller C, Willner I (2011) Angew Chem Int Ed Engl 50:295
113. Wang F, Orbach R, Willner I (2012) Chem Eur J 18:16030

Adv Biochem Eng Biotechnol (2020) 170: 59–84
DOI: 10.1007/10_2017_7
© Springer International Publishing AG 2017
Published online: 5 May 2017

Bioanalytical Application of Peroxidase-Mimicking DNAzymes: Status and Challenges

J. Kosman and B. Juskowiak

Abstract DNAzymes with peroxidase-mimicking activity are a new class of catalytically active DNA molecules. This system is formed as a complex of hemin and a G-quadruplex structure created by oligonucleotides rich in guanine. Considering catalytic activity, this DNAzyme mimics horseradish peroxidase, the enzyme most commonly used for signal generation in bioassays. Because DNAzymes exhibit many advantages over protein enzymes (thermal stability, easy and cheap synthesis and purification) they can successfully replace HRP in bioanalytical applications. HRP-like DNAzymes have been applied in the detection of several DNA sequences. Many amplification techniques have been conjugated with DNAzyme systems, resulting in ultrasensitive bioassays. On the other hand, the combination of aptamers and DNAzymes has led to the development of aptazymes for specific targets. An up-to-date summary of the most interesting DNAzyme-based assays is presented here. The elaborated systems can be used in medical diagnosis or chemical and biological studies.

Keywords DNAzyme, G-quadruplex, Hemin, Peroxidase

Contents

J. Kosman (✉) and B. Juskowiak
Laboratory of Bioanalytical Chemistry, Faculty of Chemistry, Adam Mickiewicz University,
Poznan, Poland
e-mail: kosman@amu.edu.pl

Abbreviations

ABTS	2,2′-Azino-bis(3-ethylbenzothiazoline-6-sulfonic acid)
ATP	Adenosine triphosphate
BCL-2	Regulator protein of apoptosis
C-MYC	Regulatory gene of transcription factor
CRET	Chemiluminescence resonance energy transfer
ELISA	Enzyme-linked immunosorbent assay
EXPAR	Exponential amplification reaction
G4	G-quadruplex
HCR	Hybridization chain reaction
HRP	Horseradish peroxidase
LAMP	Loop-mediated amplification
NESA	Nicking endonuclease signal amplification
PCR	Polymerase chain reaction
RCA	Rolling circle amplification
RET	Receptor tyrosine kinase
SDA	Strand displacement amplification
SELEX	Systematic evolution of ligand by exponential enrichment
SNP	Single nucleotide polymorphism
TBA	Thrombin binding aptamer
TdT	Terminal deoxynucleotidyl transferase
TMB	3,3′,5,5′-Tetramethylbenzidine
VEGF	Vascular endothelial growth factor

1   Introduction

One of the major directions in the development of medical and environmental biosensing methods is the application of DNA probes. This approach allows one to reduce costs, avoid sophisticated instrumentation, and trained staff as in routinely used PCR and ELISA techniques, which require the use of protein enzymes and antibodies. The field of DNA biosensors has progressed by utilizing their advantages over proteins, namely thermal stability, low-cost synthesis and purification, better resistance to hydrolysis, and the ability to form complicated nanostructures through hybridization between complementary strands. The other feature that makes DNA an ideal biosensor material is the ability to bind ligands selectively.

Such specific oligonucleotides are called aptamers and are developed through the SELEX approach [1–3]. Briefly, a library of various nucleic acid sequences (RNA or DNA) flanked by primer recognizing domains are exposed to the ligand. DNA/ligand complexes are then separated and amplified by PCR for the next step of selection. This process is repeated several times to obtain sequences with the highest affinity to a target molecule. Thanks to a selective binding of chosen molecules, aptamers can successfully replace protein antibodies and have been used in the development of many bioassays [4–6]. The discovery of aptamers supported the RNA-world hypothesis that the RNA molecule served as both genetic information carrier and enzyme long before the emergence of DNA and proteins [7, 8]. This hypothesis is also supported by the existence of catalytic nucleic acids. The first ribozymes were discovered by Cech and Altman in the early 1980s [9, 10]. Subsequently the repertoire of available catalyzing nucleic acids has been expanded by artificially obtained deoxyribozymes [11, 12].

One of deoxyribozymes that attracts great attention is peroxidase-mimicking DNAzyme [13]. This system was discovered when Travascio and co-workers searched for aptamers for porphyrins including hemin [14]. They identified DNA oligonucleotide, which selectively bound the hemin molecule and enhanced its peroxidase activity. This guanine-rich oligonucleotide named PS2.M (5″-GTGGGTAGGGCGGGTTGG-3′) formed a tertiary structure called G-quadruplex. Peroxidase-mimicking DNAzymes catalyze the reaction between hydrogen peroxide and organic substrates. A similar reaction is commonly used in ELISA and other molecular biology techniques supported with protein enzyme HRP. The advantages of DNAzymes over protein enzymes led to the extensive development of bioanalytical methods based on this DNAzyme. The objective of this review is to summarize knowledge about peroxidase-mimicking DNAzyme and factors influencing its activity and to present bioanalytical assays which were developed using this system as a signal amplification mechanism.

2 State of the Art

2.1 Structure and Characteristics

G-quadruplexes are formed by oligonucleotides rich in guanine that form four-stranded structures, stabilized by Hoogsteen hydrogen bonds between four guanines, creating planar G-tetrads [15]. G-quadruplexes are formed when at least two tetrads are stacked on each other (via $\pi–\pi$ interactions). Additionally, G-quadruplexes are stabilized by coordinated metal (usually K^+ or Na^+) or NH_4^+ cations interacting with O6 atoms of guanine residues. The cation is located in a central channel of the G-quadruplex. Depending on the number of DNA molecules, we can distinguish tetramolecular, bimolecular, and unimolecular structures. The direction of DNA strands also has significant impact on the G-quadruplex features and formed topologies can be characterized as parallel, antiparallel, or mixed. G-quadruplex structures have the ability to form many structures and it was found that some sequences can form different structures in the presence of potassium or sodium cations, or even form a mixture of several topologies at the same time [16]. The biological importance of G-quadruplex topologies is because bioinformatics analysis of the human genome found over 370,000 sites capable of forming G-quadruplex assemblies [17]. These sequences tend to occur more often in promotor regions of oncogenes, for example, C-MYC [18], BCL-2 [19], VEGF [20], and RET [21]. G-quadruplexes were also found to form at the ends of chromosomes called telomeres [22]. Another interesting feature of these structures is their ability to bind molecules selectively. It appears that many aptamers are G-quadruplex-forming sequences, similar to aptamers for thrombin [23], HIV-1 integrase [24], ATP [25], and nucleolin [26]. All these factors have contributed to the intensification of research in this field for the practical exploitation of G-quadruplexes in medicine and diagnostics [27].

Hemin (Fe(III)-protoporphyrin IX) is a cofactor of various oxidoreductive enzymes such as catalases and peroxidases. Besides being an active center of enzymes, hemin itself possesses some catalytic activity [28] and can catalyze hydrogen peroxide decomposition [29]. The peroxidase activity of hemin can be enhanced by binding with DNA aptamer. Researchers hypothesized that binding by other molecules may significantly enhance hemin activity. The first attempt included application of organic polymers but such complexes barely acted as monooxygenases [30]. The development of hemin aptamer initiated the research on peroxidase-mimicking DNAzymes. G-quadruplex-forming oligonucleotides bind hemin through end-stacking interactions (Fig. 1). The planar molecule of

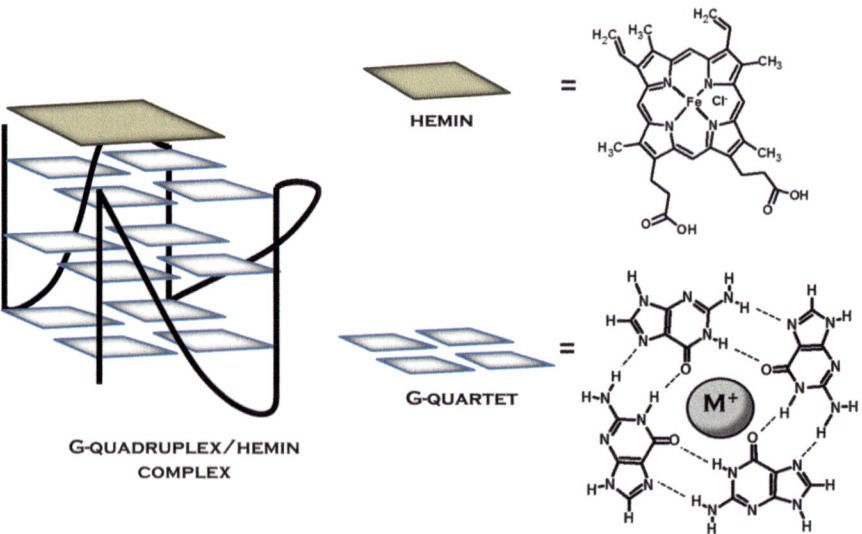

Fig. 1 Scheme of peroxidase-mimicking G-quadruplex/hemin DNAzyme

porphyrin interacts with aromatic rings of guanine bases. This interaction is more favorable for parallel G-quadruplexes because the upper G-tetrad is exposed for hemin binding. In the case of antiparallel G4, the loops overhanging the external G-tetrad create steric hindrance for hemin binding. It has been shown that DNAzymes based on parallel G-quadruplexes exhibit higher peroxidase activity [31–33]. This confirmed the relationship between the binding constant of G4/hemin complex and the catalytic activity of DNAzyme. In the first reports there was a discrepancy in the characterization of PS2.M topology. Sen and co-workers first proposed the antiparallel model of G-quadruplex structure (PS2.M in the presence of K^+) with the possibility of co-existence of more than one conformation [34, 35]. Later, Majhi and Shafer postulated that PS2.M formed rather multistranded parallel G-quadruplexes [36]. Currently it is believed that this oligo-nucleotide forms a mixture of co-existing structures consistent with the CD spectrum as well as enzymatic activity of the PS2.M/hemin system [37, 38]. Further experiments showed that DNAzyme based on the CatG4 sequence (5'-TGGGTAGGGCGGGTTGGGAAA-3'), which formed only parallel topology, possessed even higher peroxidase activity than PS2.M-based system [31]. All these reports were in good agreement concerning the influence of G-quadruplex topology on DNAzyme efficiency. This factor is crucial for the proper selection of the DNA sequence in the development of bioanalytical assays with DNAzymes.

Development of new probes and bioassays must be preceded by characterization of G-quadruplex and G4/hemin complexes. A variety of analytical techniques are useful in characterization of such systems. The G-quadruplexes are typically characterized by determination of their stability (melting profiles from UV-Vis, CD, or fluorimetry) and topology (CD spectra) [39, 40]. Structural studies at a molecular level were only undertaken by Yamamoto and co-workers [41]. They proved that hemin binds to the external G-quartet of G4. Their study, however, included only multistranded G-quadruplexes. The studies of the G-quadruplex/hemin structure by NMR or X-ray crystallography on unimolecular G4 has not yet been undertaken. The G-quadruplex/hemin complex is typically characterized by binding constant and enzymatic efficiency.

2.2 Activity and Conditions

Peroxidase-mimicking DNAzyme catalyzes the reduction of hydrogen peroxide and oxidizes an organic substrate. The cycle of classical peroxidation reaction starts from the H_2O_2 withdrawing two electrons from Fe^{III} with the creation of Compound I. In the next step, Compound I withdraws one electron from the substrate and forms Compound II. The last step involves withdrawing another electron from the substrate and the regeneration of enzyme [42]. The mechanism of G4/hemin action is believed to be a two-electron oxidation with direct transfer of oxygen from compound I to the substrate [43]. Through the selection of adequate substrates, peroxidase-mimicking DNAzyme systems can be used as a source of analytical signal. Typically, substrates are chosen to obtain products exhibiting chemiluminescence (luminol), fluorescence (Amplex Red, Thiamine, MNBDH), color change (ABTS, TMB), or generation of electrochemical signals. Shangguan et al. showed that DNAzyme had a broader substrate specificity than HRP. This is because the hemin active center is more exposed in a G4/hemin complex than in a protein enzyme [44]. Their research also provided insight into the inactivation process of DNAzyme which resulted from hemin degradation by H_2O_2.

The activity of DNAzymes is affected by several factors. The most important element is a sequence of DNA oligonucleotides. As mentioned earlier, DNAzymes based on parallel G-quadruplexes tend to exhibit higher activity than antiparallel G4 systems. The topology of G-quadruplexs is connected with the type of cation present in the solution. First reports suggested that potassium cation was essential for peroxidase activity of DNAzyme [13]. This assumption was connected with G-quadruplex capability to form predominantly parallel G-quadruplex in the presence of K^+, in contrast to antiparallel G-quadruplex stabilized by sodium cation [16]. Later reports points out, however, that the highest activity is observed when ammonium cation is present [45, 46]. The enhancement effect of ammonium ion on DNAzyme activity was attributed to the higher resistance of NH_4^+-stabilized G-quadruplex to degradation by radical reaction products. Another factor that plays a crucial role in DNAzyme activity is a choice of buffer composition. Generally, nitrogenous buffers supported the activity of DNAzyme, whereas oxyanion buffers had an inhibitory effect [13]. Another important aspect is the choice of the substrate for peroxidation reactions. Nakayama and Sintim have also demonstrated that DNAzymes based on various sequences differ in catalytic activity with reference to different substrates. These reports demonstrated the significance of G4 sequence, substrate nature, and cationic condition selection for the design of efficient sensing systems based on DNAzymes.

To enhance DNAzyme activity, researchers tried to find substances (cofactors) that could improve its efficiency. Kong et al. showed that ATP enhanced catalytic activity and inhibited $ABTS^{·+}$ disproportion [47, 48]. Stefan and co-workers further studied the effect of various nucleotides on chosen DNAzymes and found that all nucleotides had a positive effect on catalytic activity, with ATP being the best enhancer [49]. Another substance that had a positive effect on the catalytic activity of peroxidase-mimicking DNAzyme was spermine, which showed activity enhancement in systems with various substrates [50]. Spermine's enhancing effect arose from the protection of hemin against degradation by H_2O_2 as well as from the more compact structure of G4 and creation of specific microenvironment for peroxidase activity.

Sen and co-workers recently reported that G4/hemin complex was also able to catalyze two-electron oxidation reactions [51]. They used thioanisole, indole, and styrene as substrates of an oxidation reaction. By applying labeled $H_2^{18}O_2$ they showed that oxygen derived from hydrogen peroxide was transferred to the substrates. The ability of G4/hemin complex to catalyze two-electron oxidation is interesting in the light of industrial application. This finding was later exploited to show that G4/hemin DNAzyme was able to mimic not only horseradish peroxidase but also NADH oxidase and NADH peroxidase [52, 53]. The expansion of the DNAzyme repertoire by new activities provides new possibilities in bioassay development.

2.3 Modifications

G4/hemin DNAzymes have been shown to be excellent catalysts, ideal for bioanalytical approaches. However, studies on improving their performance and catalytic properties are still ongoing. The main goal is to reduce some drawbacks, mainly dissociation of G4/hemin complex (important in heterogeneous applications) and to enhance their activity even more. To shed more light on the influence of DNA sequence on catalytic activity, the researchers studied DNAzymes with additional nucleotides. Kong et al. investigated the addition of T nucleotides to $T_nG_4T_m$ sequences and noticed that additional T nucleotides on the 5' end enhanced DNAzyme activity, whereas elongation of the 3' end by thymidine had little to no effect [32]. Later on, Shangguan developed DNAzyme with d(CCC) sequence flanking both ends of the G4 oligonucleotide [54]. His results suggested that such a modification had a stabilizing effect on G-quadruplex/hemin complexes and an enhancing effect on DNAzyme activity. Li and co-workers have recently studied DNAzymes with adenine at the 3' end of the oligonucleotide [55]. Adenine at the 3' end functioned as a general acid-base catalyst. In their paper the authors claimed that adenine acted as a distal ligand of hemin, similar to histidine in horseradish peroxidase. The superiority of CatG4 DNAzyme over other G4 sequences was pointed out. CatG4 exhibits very high catalytic activity and also possesses adenine residue at the 3' end. All these studies shed more light on the DNA sequence influence on activity and provided knowledge for designing DNAzymes with the best features.

In 2005 Niemeyer et al. published a paper on covalent attachment of hemin to DNA oligonucleotide [56]. Their research focused, however, on the reconstruction of protein enzyme on solid support. Hemin was conjugated to DNA oligonucleotide using amine coupling. This approach was later used by Sintim and co-workers for creation of the G4–hemin conjugate with catalytic activity [57]. They showed that conjugation of hemin to DNA sequence significantly enhanced catalytic properties, especially in relation to sequences that normally possess very low activity (e.g., telomeric sequence GGG(TTAGGG)$_3$). The study on G4–hemin conjugates was continued by other groups and provided more information on the properties of covalent DNAzymes such as the independence of catalytic activity on the presence of cations [58–60]. The covalent attachment of hemin is a very promising approach because it offers limitation of the blank signal that is especially advantageous in heterogeneous sensing formats.

Other modifications of peroxidase-mimicking DNAzymes involved attachment of additional molecules to the end of DNA oligonucleotides. Xiao and co-workers used cationic peptide for enhancement of peroxidase and oxidase activity of

DNAzymes [61]. The activity of DNAzyme–peptide conjugate was fourfold higher than that for unmodified DNAzyme. To increase solubility of DNAzymes in organic solvents, DNAzyme was conjugated with polyethylene glycol [62]. Such biocatalytic constructs can be used for industrial applications in which many chemical processes are carried out in organic solvents. Lastly, Li and co-workers investigated the activity of DNAzymes based on analogues of DNA [63]. Their study revealed that the highest activity was achieved for 2′-OMe-RNA catalytic units. The higher activity was attributed to the *anti* conformation of 2′-OMe-RNA, which is more favorable for the formation of parallel G-quadruplexes exhibiting superior enzymatic activity. The presented modified DNAzymes can also contribute to the development of more sensitive and versatile biosensing systems.

3 Bioanalytical Applications

3.1 Detection of DNA and microRNA

The greatest increase in the development of biosystems based on DNAzymes is observed for bioassays for nucleic acids. This is connected with the availability of many amplification methods and sophisticated nucleic acid constructs. The invention of new methods for DNA detection is important because DNA can be easily detected as a biomarker for diseases such as genetic disorders and pathogen-related diseases. Great developments have been observed in assays aimed at microRNA detection. MicroRNAs are short oligonucleotides which regulate protein translation, especially in cell proliferation, differentiation, and apoptosis [64]. Such strong interest in these molecules is caused by their availability in body fluids, their small size, and their connection to many diseases. The simplest but sensitive homogeneous assay for nucleic acid is based on molecular beacons [65–72]. In this approach, a DNA probe is divided into three domains (Fig. 2a). The hairpin loop is complementary to the nucleic acid analyte. Half of the DNAzyme-forming sequence is locked by a complementary sequence. When the analyte is present it forms a duplex with the loop domain, opening the hairpin structure, and freeing the DNAzyme-forming sequence. After potassium and hemin addition, DNAzyme catalyzes substrate oxidation. This approach was also used in an inhibition approach where two molecular beacon probes could form split DNAzyme in the absence of DNA analyte [73]. The assays based on molecular beacons and DNAzymes were used in the detection of *Pseudostellaria heterophylla* [73], microRNA [72], and various artificial DNA sequences. It should be noted, however, that this method can be easily transferred to the detection of any chosen DNA by changing the probe sequence. Another simple approach involves the use of split DNAzymes [74–78]. This method is based on creation of DNAzyme by two separate DNA probes after hybridization with a complementary strand (Fig. 2b). This approach can involve two equal probes, where the DNAzyme-forming

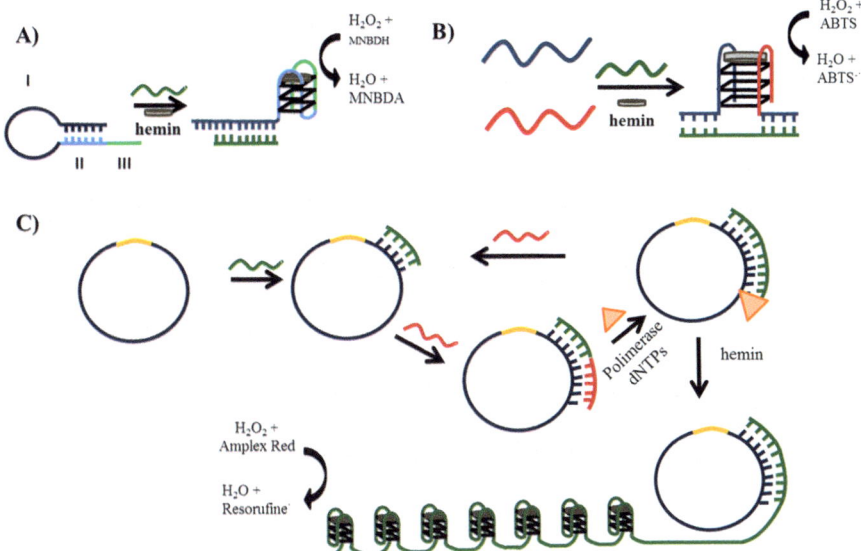

Fig. 2 Scheme of DNA detection formats. (**a**) Molecular beacon probe. (**b**) Split DNAzymes. (**c**) RCA-DNAzyme coupled approach

sequence is cut in half or asymmetrically. Researchers developed bioassays based on split DNAzymes for the detection of a single nucleotide polymorphism (SNP), mutations in bacteria [76], or genetically modified food [78].

The evolution of molecular biology techniques has enabled the development of sophisticated DNA assays. In particular, the combination of protein enzymes and DNAzymes gave promising results. Such an approach allows for the use of two amplification techniques in the same assay. One example of this approach is the application of rolling circle amplification (RCA) and G4/hemin DNAzymes [79–84]. RCA is a replication technique which allows for the synthesis of many linear copies of circular DNA or RNA. In this method, DNA primer binds to a circular probe and a double stranded DNA tract is then recognized by DNA polymerase, which synthesizes DNA amplicon with a sequence complementary to the circular probe. As a result, many replicas of DNAzyme-forming sequences are synthesized and, after hemin and potassium cation addition, the G4/hemin system catalyzes substrate oxidation (Fig. 2c). The application of two amplification methods allows one to detect DNA at fM or even aM scale. Several modifications of the RCA method were developed to enhance system sensitivity. Wen et al. proposed the application of triggered polycatenated DNA scaffolds [82]. For this purpose they developed a DNA capture probe immobilized on magnetic beads. The binding of DNA analyte allows subsequent binding of a padlock probe. After ligase operation and additional probes addition, the polycatenated scaffolds are formed, which are then amplified using RCA. As a result, thousands of DNAzyme-forming sequence copies are synthesized. The application of streptavidin-modified magnetic beads and biotinylated primers allows for the separation of RCA products and hence reduces the signal from hemin in a reference sample [82]. Another interesting modification of the RCA method was developed by Wen and co-workers [84]. In this approach, the RCA product is subjected to nicking endonuclease signal amplification (NESA). The double amplification of DNA probe is continued by DNAzyme formation and substrate oxidation. Application of these amplification methods resulted in microRNA detection limit of 2 aM [84].

Numerous methods for DNA detection are based on the utilization of polymerase enzyme in an isothermal polymerization format. Isothermal methods are very desirable because, unlike PCR, they do not require temperature changes and sophisticated thermocyclers. Generally, these methods involve cyclic formation of probes through enzymatic activity of polymerase, which can also be coupled with endonuclease activity [85–98]. The isothermal amplification techniques can be arranged in several approaches, such as exponential isothermal amplification reaction (EXPAR), strand displacement amplification (SDA), hybridization chain reaction (HCR), or loop-mediated amplification (LAMP). Shen and co-workers designed system that involved hairpin probes which, after hybridization with DNA analyte, can be subjected to elongation by Taq polymerase (Fig. 3a) [90]. After dehybridization, the DNA analyte can be reused to hybridize with the next hairpin probe in the subsequent cycle. The oligonucleotide synthesized by DNA polymerase consists of a G-quadruplex-forming sequence and can form DNAzyme that catalyzes peroxidase reaction. The interesting approach was also

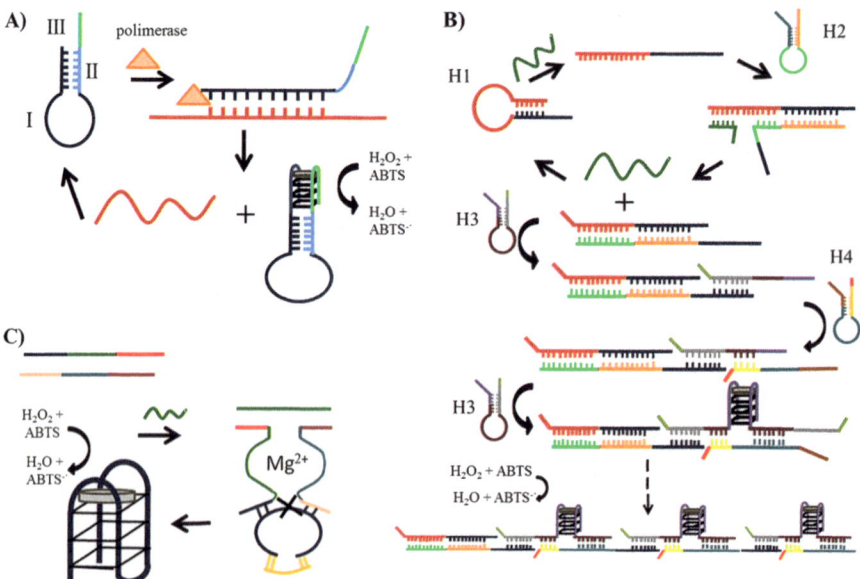

Fig. 3 Illustration of DNA detection methods. (a) Isothermal enrichment amplification. (b) Hybridization strand displacement. (c) Cooperation of peroxidase-mimicking DNAzyme and Mg^{2+}-DNAzyme

presented by Xu et al., which was based on the usage of polymerase and endonuclease enzymes in EXPAR [97]. The first stage involved generation of a specific site in probe/analyte duplex that was recognized by endonuclease. Next, the duplex was subjected to polymerase elongation and creation of another recognition site for endonuclease activity, followed by releasing synthesized primers for a second DNA template. This probe also involved sites recognized by endonuclease after duplex formation. The final step consists of releasing the DNAzyme-forming sequence through endonuclease activity. The combination of elaborated polymerase and endonuclease activity resulted in amplification of DNAzyme copies resulting in the detection of targeted microRNA at a 10 fM concentration level. The interesting approach of coupled catalyzing hairpin assembly and hybridization chain reaction (HCR) was investigated by Wu and co-workers [98]. This method does not require enzyme activity, and amplification of the DNAzyme units is achieved by cyclic hybridization of hairpin probes and hybridization of templates (Fig. 3b). The binding of a microRNA target by a hairpin probe resulted in the formation of a duplex structure. In the next step, the second hairpin probe binds to duplex and elongates it. The DNAzyme-forming unit is released but elongation of the duplex structure proceeds. Many of the DNAzyme units are formed on this construct created by one microRNA target molecule. This method allows for the detection of microRNA on a 7.4 fM scale. The method based on dual DNAzymes activity was also investigated [99]. In this system the peroxidase-mimicking DNAzyme is coupled with Mg^{2+}-dependent DNAzyme (Fig. 3c). Several DNA probes are designed to form a nicking site only after DNA analyte hybridization. The Mg^{2+}-dependent DNAzyme catalyzes DNA nicking that releases the peroxidase-mimicking DNAzyme-forming probes.

Developments in nanotechnology techniques have opened up new possibilities for designing innovative DNAzyme systems. The highly advantageous properties of nanomaterials include plasmon-enhanced optical properties of gold or silver nanoparticles, highly fluorescent quantum dots, or easy separation using magnetic particles. Generally, these new methods also employ detection constructs and amplification techniques as described above. The nanoparticles are used for signal generation and creation of a heterogeneous environment in the solution that facilitates analyte separation. Quantum dots exhibit high fluorescence intensity, which can be used for designing detection systems based on chemiluminescence resonance energy transfer (CRET) [100–102]. Liu and co-workers developed a system, in which DNA oligonucleotide was conjugated with CdSe/ZnS quantum dots. DNAzyme-catalyzed luminol oxidation and generated chemiluminescence was transferred to quantum dots, which emitted fluorescence signals. The amplification effect can also be obtained using AuNPs, which are modified with two different oligonucleotide probes [103, 104]. One was complementary to the DNA analyte (capture probe) and the other was able to form DNAzymes on the AuNP (reporting probe). Hybridization of the reporting probe with the part of the analyte sequence which, in a previous step, was immobilized on capture probe resulted in the formation of many DNAzymes on the basis of one DNA target molecule.

3.2   Detection of Proteins/Enzymes

Peroxidase-mimicking DNAzymes also found application in protein and enzyme activity detection. A significant number of researchers developed methods for the detection of thrombin [105–116]. It was established that thrombin aptamer TBA (5′-GGT TGG TGT GGT TGG-3′) adopted the G-quadruplex structure [117]. Hemin has a very low binding affinity to TBA G-quadruplex, so the activity of formed DNAzyme is also very low. However, binding of thrombin by aptamer causes an increase in affinity toward hemin and, as a result, enhances peroxidase activity of DNAzyme. Exploiting this phenomenon has led to the development of thrombin biosensors, which can be built as homogeneous or heterogeneous systems. The simplest approach was introduced by Li et al. and was based on formation of DNAzyme by TBA oligonucleotide after binding of thrombin [105]. This method allowed one to detect thrombin at a 20 nM concentration level. Heterogeneous systems for thrombin detection were obtained using a DNA probe attached to the surface of the solid support (Fig. 4a). This improvement led to increasing of aptasensor performance by lowering the detection limit to 100 pM of thrombin [106]. The other aptasensors for thrombin included TBA aptamer connected with other DNAzyme-forming sequences [109]. The interesting feature of thrombin is that this molecule possesses two aptamer-binding sites. This can be utilized in the

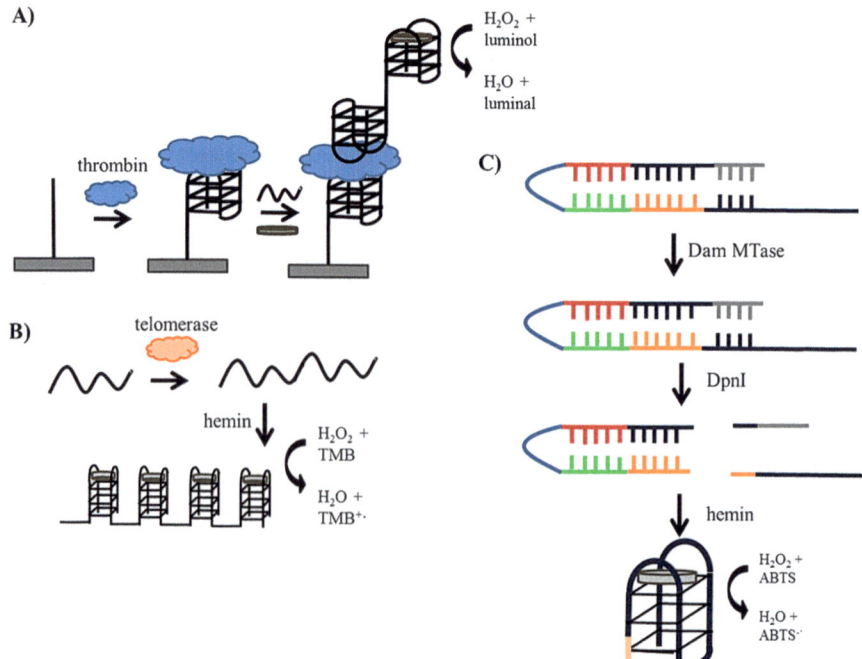

Fig. 4 Illustration of protein detection approaches. (**a**) Heterogeneous method of thrombin detection. (**b**) Homogeneous telomerase assay. (**c**) Methyltransferase activity detection method

development of an assay in which one aptamer is used as a capture probe and the second functions as a reporter DNAzyme [107].

The systems based on peroxidase-mimicking DNAzymes were also used in the development of assays for protein cancer biomarkers [118–130]. One example of such a biomarker is telomerase. This enzyme is a reverse transcriptase, which adds (TTAGGG) repeats to the ends of chromosomes. However, this enzyme is only active in stem cells and cancer cells. Telomerase is present in more than 80% of all cancers and is therefore a very good cancer biomarker [131]. The telomeric DNA is able to form G-quadruplex structures, although its DNAzyme possesses very low activity. The first methods of telomerase detection based on G4/hemin DNAzyme used the external probe of PS2.M DNAzyme [65, 118, 119]. Willner's group used three approaches: molecular beacon, immobilization of the capture probe on the surface, and addition of DNAzyme probe and AuNPs modified with capture probe and DNAzyme sequence (Fig. 4b). This methods allowed for the detection of telomerase with a detection limit of 1,000 HeLa cells. The direct method of telomerase activity determination based on peroxidase activity of DNAzymes formed on telomeric strands was also developed [122, 130]. The optimization of measurement conditions allowed for the development of telomerase assays with a detection limit of 200 HeLa cells.

Activity of other enzymes has been also investigated using methods based on deoxyribozymes. So far, detection methods have been developed for methyltransferase [132–134], DNA glycosylase [135], DNA ligase [136, 137], caspase 3 [138], TdT [139], T4 polynucleotide kinase [137, 140], and S1 nuclease [141]. The major principle of these methods was generation or destruction of DNAzymes as a result of catalytic activity of target enzymes. A typical example of such a system is the assay for methyltransferase shown in Fig. 4c [132]. The hairpin probe consisted of five domains: domain I was able to form DNAzyme but was immobilized by partially complementary domain III. DNA methyltransferase recognized the site in the hairpin sequence and generated methylated DNA duplex, which in the next step served as a substrate for endonuclease DpnI. As a result of both enzyme activities, the G-quadruplex-forming sequence was released and DNAzyme was able to catalyze the peroxidase reaction (Fig. 4c). The sensitivities of the enzyme assays varied from 0.06 to 0.00013 U/ml.

3.3 Detection of Metals

The formation of the G-quadruplex structure is a cation-dependent process. The formation of active DNAzyme proceeds with a better yield in the presence of potassium cations than other metal cations. On this basis, many potassium bio-assays have been developed [103, 142–144]. The simplest method involved forma-tion of G-quadruplex by DNA oligonucleotide only in the presence of potassium (Fig. 5a) [142–144]. Willner and co-workers developed the electrochemical assay for potassium based on AuNPs modified with a DNAzyme-forming sequence. This

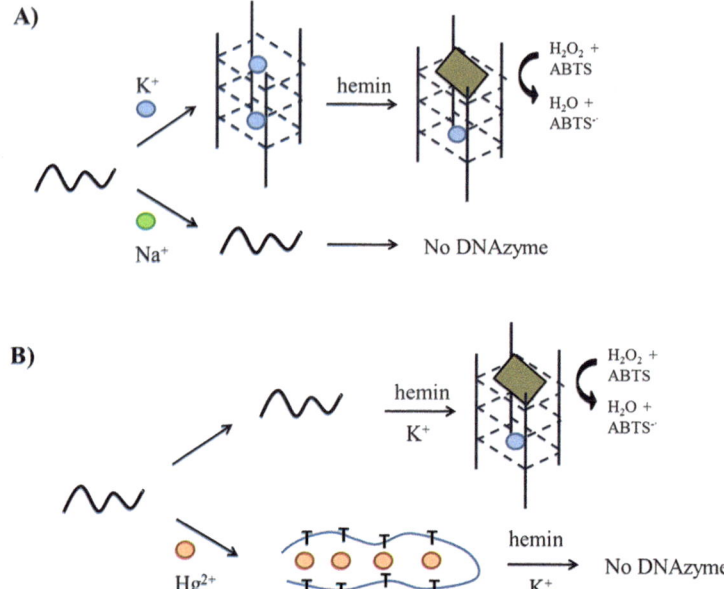

Fig. 5 Illustration of detection methods for metal cations. (**a**) Potassium. (**b**) Mercury

system was immobilized on the electrode using a DNA capture probe. The assay proved the specificity over other cations even at a higher concentration than potassium cations.

Mercury ions stabilize duplex through formation of T-Hg^{2+}-T base pairs and this property was utilized in the development of mercury bioassays [32, 145–147]. This Hg(II)-sensing system can be designed in two ways: promotion or disruption of G-quadruplex structure by mercury-stabilized base pairs. The promotion of DNAzyme generation can be carried out by the formation of intramolecular G4 structures stabilized by Hg^{2+} base pairs between four DNA molecules [32]. The other approach was used by Hao et al., who developed hairpin probes that, after Hg^{2+}-stabilized duplex formation, were also able to form G-quadruplex structures [147]. The second approach that utilized G-quadruplex disruption was chosen by the Wang group [145]. DNA oligonucleotide in the presence of mercuric ions formed self-dimers. However, when mercury was not present the G-quadruplex structure could form freely and DNAzyme could catalyze peroxidation reaction (Fig. 5b) [145]. Similar to mercury, silver ion can also stabilize duplex structures through C-Ag^{+}-C base pairs. In this case, the method also used a DNA probe that formed G-quadruplex stabilized by C-Ag^{+}-C base pairs in the duplex region [148]. Other silver detection methods utilized silver-enhanced disruption of G-quadruplex [149–152]. This disruption was not the result of base pair formation but was simply caused by the interaction of silver ions with guanines in G-quadruplex structures.

The PW17 DNAzyme derived from the PS2.M sequence forms different topologies in the presence of potassium and lead cations. Interestingly, the lead-stabilized form was unable to bind hemin molecule and did not exhibit peroxidase activity. This feature was used for development of Pb(II) bioassays [153–155]. Other methods of lead detection utilized the combination of two DNAzymes: peroxidase-mimicking and Pb^{2+}-dependent nuclease [156–158]. Pb^{2+}-dependent nicking DNAzyme catalyzed the hydrolytic cleavage of DNA and released the G-quadruplex-forming sequence with creation of HRP-mimicking DNAzyme.

3.4 Detection of Other Molecules

Combination of aptamers and DNAzymes has led to the development of aptasensors. These systems allowed for the detection of chosen molecules, for which aptamers were developed. Numerous methods were reported for adenosine [159–163]. The design of aptasensors was based on the change in aptamer conformation after adenosine binding by aptamers. Teller et al. developed a system that formed hairpin structures (Fig. 6a). The adenosine binding forced the change in the structure and release of DNAzyme unit [159]. The similar approach was presented by Chu and co-workers and involved the utilization of blocking DNA strand. In this case, the binding of target by aptamer also resulted in the release of DNA strand, which was then used for the hybridization chain reaction. The final stage of the assay included peroxidase reaction catalyzed by the DNAzyme formed [162]. The significant numbers of aptazymes were also developed for mycotoxins, such as ochratoxin A and aflatoxin B1 [164–170]. Both these toxins contaminate food and are dangerous to human health. The aptazyme design is analogous to the adenosine sensors. An interesting system was aptasensor that used CRET for signal generation

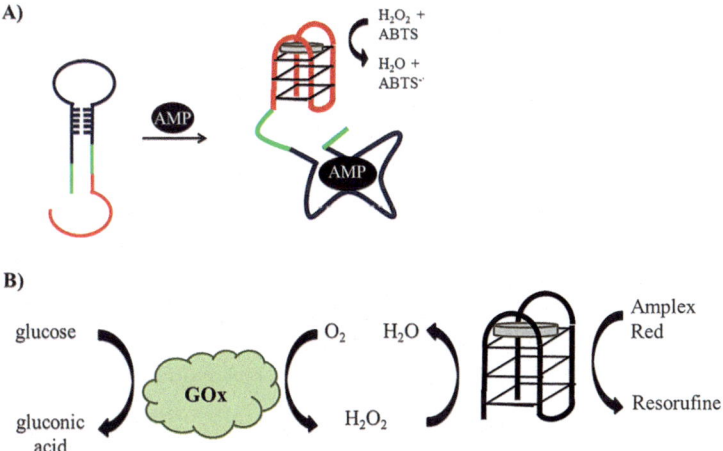

Fig. 6 Scheme of (**a**) aptazyme for adenosine and (**b**) glucose detection system

[167, 170]. The DNAzyme formed only in the presence of ochratoxin A catalyzed the luminol oxidation. The chemiluminescence was quenched by quencher on the 3' end of the aptamer. Even if the DNAzyme could form in the absence of the target, the CRET could not occur because quencher was located on the other end of DNA strand. Aptamer for aflatoxin A was based on a split aptasensor principle [168]. Three DNA molecules in the absence of the target molecule formed split DNAzyme stabilized by forming duplex with one long DNA strand. In the presence of toxin the aptamer was associated with the target molecule and could not stabilize split DNAzyme, hence blocking peroxidase activity. Aptazymes for cocaine were also developed [171, 172] as well as for coralyne [173, 174], cytokine [175, 176], and daunomycin [177].

By combination of DNAzyme activity and glucose oxidase it was possible to develop glucose sensors [112, 178–180]. Glucose oxidase catalyzes glucose oxidation to gluconic acid with generation of hydrogen peroxide, which can then be used for peroxidase reaction catalyzed by DNAzyme (Fig. 6b). The similar sensor was developed for cholesterol using cholesterol oxidase [181]. An interesting application of DNAzymes is monitoring the antioxidant properties of studied molecules. Such an assay was developed by Wang et al. [182]. DNAzyme catalyzed ABTS oxidation to cation radical ABTS$^-$·. Substances with antioxidant properties captured radicals that resulted in a change of solution color.

4 Outlook

Despite nearly 20 years of research on the G4/hemin DNAzymes, there is still room for improvement of their activity and usability. Great potential can be seen for modified DNAzymes that can allow for expansion of possible applications. Both the use of DNA analogues and covalent attachment of hemin molecules can increase DNAzyme activity. Very promising is also the attachment of PEG that can enable development of sensors to perform in organic solutions for industrial applications.

Following the discovery of the peroxidase-mimicking DNAzyme in 1998, this system has been comprehensively used in the development of bioassays. The availability of many amplification techniques led to the introduction of ultrasensitive assays. DNAzymes were also shown to exhibit many advantages over protein enzymes, such as thermal stability and easy and cheap synthesis and purification. Despite this, DNAzyme-based biosensors have not managed to replace routinely used protein assays. The scientific community recognizes the great potential of this system, as evidenced by the large number of papers on this topic. However, there is still a need for development of kits and ready-made sensors.

References

1. Oliphant AR, Brandl CJ, Struhl K (1989) Defining the sequence specificity of DNA-binding proteins by selecting binding sites from random-sequence oligonucleotides: analysis of yeast GCN4 proteins. Mol Cell Biol 9:2944–2949
2. Ellington ED, Szostak JW (1990) In vitro selection of RNA molecules that bind specific ligands. Nature 346:818–822
3. Tuerk C, Gold L (1990) Systematic evolution of ligands by exponential enrichment: RNA ligands to bacteriophage T4 DNA polymerase. Science 249:505–510
4. Kang KN, Lee YS (2013) RNA aptamers: a review of recent trends and applications. Adv Biochem Eng Biotechnol 131:153–169
5. Hong P, Li W, Li J (2012) Application of sensors in clinical diagnostics. Sensors 12 (2):1181–1193
6. Citartan M, Gopinath SCB, Tominaga J, Tan SC, Tang TH (2012) Assays for aptamer-based platforms. Biosens Bioelectron 34(1):1–11
7. Breaker RR (2012) Riboswitches and the RNA world. Cold Spring Harb Perspect Biol 4: a003566
8. Gilbert W (1986) Origin of life: the RNA world. Nature 319(6055):618
9. Kruger K, Grabowski PJ, Zaug AJ, Sands J, Gottschling DE, Cech TR (1982) Self-splicing RNA: autoexcision and autocyclization of the ribosomal RNA intervening sequence of Tetrahymena. Cell 35:849–857
10. Guerriertakada C, Garinder K, Marsh T, Pace N, Altman S (1983) The RNA moiety of ribonuclease P is the catalytic subunit of the enzyme. Cell 35:849–857
11. Breaker RR, Ronald R (1997) DNA enzymes. Nat Biotechnol 5:427–431
12. Silverman SK, Scott K (2004) Deoxyribozymes: DNA catalysts for bioorganic chemistry. Org Biomol Chem 2(19):2701–2706
13. Travascio P, Li Y, Sen D (1998) DNA-enhanced peroxidase activity of DNA-aptamer-hemin complex. Chem Biol 5(9):505–517
14. Li Y, Geyer CR, Sed D (1996) Recognition of anionic porphyrins by DNA aptamers. Biochemistry 35:6911–6922
15. Simonsson T (2001) G-quadruplex DNA structures variations on a theme. Biol Chem 382 (4):621–628
16. Burge S, Parkinson GN, Hazel P, Todd AK, Neidle S (2006) Quadruplex DNA: sequence, topology and structure. Nucleic Acids Res 34(19):5402–5415
17. Lam EYN, Beraldi D, Tannahill D, Balasubramanian S (2013) G-quadruplexe structures are stable and detectable in human genomic DNA. Nat Commun 2013(4):1796
18. Yang C, Hurley LH (2006) Structure of the biologically relevant G-quadruplex in the c-MYC promoter. Nucleos Nucleot Nucl 25(8):951–968
19. Agrawal P, Lin C, Mathad RI, Carver M, Yang D (2014) The major G-quadruplex formed in the human BCL-2 proximal promoter adopts a parallel structure with a 13-nt loopin K$^+$ solution. J Am Chem Soc 136(5):1750–1753
20. Agrawal P, Hatzakis E, Guo K, Carver M, Yang D (2013) Solution structure of the major G-quadruplex formed in the human VEGF promoter in K+: insights into loop interactions of the parallel G-quadruplexes. Nucleic Acids Res 41(22):10584–10592
21. Tong X, Lan W, Zhang X, Wu H, Liu M, Cao C (2011) Solution structure of all parallel G-quadruplex formed by the oncogene RET promoter sequence. Nucleic Acids Res 39 (15):6753–6763
22. Biffi G, Tannahill D, McCafferty J, Balasubramanian S (2013) Quantitative visualization of DNA G-quadruplexes structures in human cells. Nat Chem 5:182–186
23. Bock LC, Griffin LC, Latham JA, Vermaas EH, Toole JJ (1992) Selection of single-stranded DNA molecules that bind and inhibit human thrombin. Nature 355:564–566

24. Phan AT, Kuryavyi V, Ma JB, Faure A, Andreola ML, Patel DJ (2005) An interlocked dimeric parallel-stranded DNA quadruplex: a potent inhibitor of HIV-1 integrase. Proc Natl Acad Sci U S A 102:634–639
25. Huizenga DE, Szostak JW (1995) A DNA aptamer that binds adenosine and ATP. Biochemistry 34:656–665
26. Ireson CR, Kelland LR (2006) Discovery and development of anticancer aptamers. Mol Cancer Ther 5:2957–2962
27. Lv L, Guo Z, Wang J, Wang E (2012) G-quadruplex as signal transducer for biorecognition events. Curr Pharm Design 18:2076–2095
28. Kelly HC, Davies MD, Mantle D, Jones P (1977) Hydroperoxidase activities of ferrihemes: heme analogues of peroxidase enzyme intermediates. Biochemistry 16:3974–3978
29. Kremer ML (1967) Decomposition of hydrogen peroxide by hemin. Trans Faraday Soc 61:1453–1459
30. Johnstone RAW, Simpson AJ, Stocks PA (1997) Porphyrins in aqueous amphiphilic polymers as peroxidase mimics. Chem Commun 23:2277–2278
31. Kong DM, Cai LL, Guo JH, Wu J, Shen HX (2008) Characterization of the G-quadruplex structure of a catalytic DNA with peroxidase activity. Biopolymers 91(5):331–339
32. Kong DM, Wu J, Wang N, Yang W, Shen HX (2009) Peroxidase activity-structure relationship of the intermolecular four-stranded G-quadruplex-hemin complexes and their application in Hg^{2+} ion detection. Talanta 80:459–465
33. Kong DM, Yang W, Wu J, Li CX, Shen HX (2010) Structure-function study of peroxidase-like G-quadruplexe-hemin complexes. Analyst 135:321–326
34. Travascio P, Witting PK, Mauk AG, Sen D (2001) The peroxidase activity of a hemin-DNA oligonucleotide complex: free radical damage to specific guanine bases of the DNA. J Am Chem Soc 123:1337–1348
35. Travascio P, Sen D, Bennet AJ (2006) DNA and RNA enzymes with peroxidase activity – an investigation into the mechanism of action. Can J Chem 84:613–619
36. Majahi PR, Shafer RH (2006) Characterization of the unusual folding pattern in a catalytically active guanine quadruplex structure. Biopolymers 82(6):558–569
37. Cheng X, Liu X, Bing T, Cao Z, Shangguan D (2009) General peroxidase activity of G-quadruplex-hemin complexes and its application on ligand screening. Biochemistry 48:7817–7823
38. Kosman J, Juskowiak B (2016) Hemin/G-quadruplex structure and activity alteration induced by magnesium cations. Int J Biol Macromol 85:555–564
39. Mergny JL, Phan AT, Lacroix L (1998) Following G-quartet formation by UV-Vis spectroscopy. FEBS Lett 435(1):74–78
40. Vorlickova M, Klejnovska I, Sagi J, Renciuk D, Bednarova K, Motlova J, Kypr J (2012) Circular dichroism and guanine quadruplexes. Methods 57(1):64–75
41. Yamamoto Y, Kinoshita M, Katahira Y, Shimizu H, Di Y, Shibata T, Tai H, Suzuki A, Neya S (2015) Characterization of heme-DNA complexes composed of some chemically modified hemes and parallel G-quadruplex DNAs. Biochemistry 54(49):7168–7177
42. Everse J, Johnson M, Marini M (1994) Methods in enzymology vol. 231, hemoglobins, Part B, biochemical and analytical methods. Academic Press, San Diego, pp. 547–561
43. Poon L, Methot S, Morabi-Pazooki W, Pio F, Bennet A, Sen D (2011) Guanine-rich RNAs and DNAs that bind heme robustly catalyze oxygen transfer reactions. J Am Chem Soc 11:1877–1884
44. Yang X, Fang C, Mei H, Chang T, Cao Z, Shangguan D (2011) Characterization of G-quadruplex/hemin peroxidase: substrate specificity and inactivation kinetics. Chem Eur J 17:14475–14484
45. Nakayama S, Sintim HO (2009) Colorimetric split G-quadruplexes probes for nucleic acid sensing: improving reconstituted DNAzyme's catalytic efficiency vie probe remodeling. J Am Chem Soc 131:10320–10333

46. Nakayama S, Sintim HO (2012) Investigating the interactions between cations, peroxidation substrates and G-quadruplex topology in DNAzyme peroxidation reactions using statistical testing. Anal Chim Acta 747:1–6

47. Kong DM, Xu J, Shen HX (2010) Positivie effects of ATP on G-quadruplex-hemin DNAzyme-mediated reactions. Anal Chem 82:6148–6153

48. Jia SM, Liu XF, Kong DM, Shen XH (2012) A simple, post-additional antioxidant capacity assay using adenosine triphosphate-stabilized 2,2′-azinobis(3-ethylbenzothiazoline)-6-sulfonic acid (ABTS) radical cation in a G-quadruplex DNAzyme catalyzed ABTS–H2O2 system. Biosens Bioelectron 35:407–412

49. Stefan L, Denat F, Monchaud D (2012) Insights into how nucleotide supplements enhance the peroxidase-mimicking DNAzyme activity of the G-quadruplex/hemin system. Nucleic Acids Res 40(17):8759–8772

50. Qi C, Zhang N, Yan J, Liu X, Bing T, Mei H, Shangguan D (2014) Activity enhancement of G-quadruplex/hemin DNAzyme spermine. RSC Adv 4:1441–1448

51. Poon LCH, Methot SP, Morabi-Pazooki W, Pio F, Bennet AJ, Sen D (2011) Guanine-rich RNAs and DNAs that bind hemin robustly catalyse oxygen transfer reactions. J Am Chem Soc 133:1877–1884

52. Golub E, Freeman R, Willner I (2011) A hemin/G-quadruplex acts as an NADH oxidase and NADH peroxidase mimicking DNAzyme. Angew Chem Int Ed 50:1–6

53. Yuan Y, Yuan R, Chai Y, Zhuo Y, Ye X, Gan X, Bai L (2012) Hemin/G-quadruplex simultaneously acts as NADH oxidase and HRP-mimicking DNAzyme for simple, sensitive pseudobienzyme electrochemical detection of thrombin. Chem Commun 48:4621–4623

54. Chang T, Gong H, Ding P, Liu X, Li W, Bing T, Cao Z, Shangguan D (2016) Activity ehancment of G-quadruplex/hemin DNAzyme by Flanking d(CCC). Chem Eur J 22:4015–4021

55. Li W, Li Y, Liu Z, Lin B, Yi H, Xu F, Nie Z, Yao Z (2016) Insight into G-quadruplex-hemin DNAzyme/RNAzyme adjacent adenine as the intermolecular species for remarkable enhancement of enzymatic activity. Nucleic Acids Res 44(15):7373–7384

56. Fruk L, Niemeyer CM (2006) Covalent hemin-DNA adducts for generating a novel class of artificial heme enzymes. Angew Chem Int Ed 44:2603–2606

57. Nakayama S, Wang J, Sintim HO (2011) DNA-based peroxidation catalyst-what is the exact role of topology on catalysis and is there a special binding site for catalysis? Chem Eur J 17:5691–5698

58. Gribas AV, Korolev SP, Zatsepin TS, Gottikh MB, Sakharov IY (2015) Structure-activity relationship study for design of highly activr covalent peroxidase-mimicking DNAzyme. RSC Adv 5:51672–51677

59. Gribas AV, Korolev SP, Zatsepin TS, Gottikh MB, Sakharov IY (2016) Suicide inactivation of covalent peroxidase-mimicking DNAzyme with hydrogen peroxide and its protection by a reductant substrate. Talanta 155:212–215

60. Wang Z, Zhao J, Bao J, Dai Z (2016) Construction of metal-ion-free G-quadruplex-hemin DNAzyme and its application in S1 nuclease detection. ACS Appl Mater Interfaces 8:827–833

61. Xiao L, Zhou Z, Feng M, Tong A, Xiang Y (2016) Cationic peptide conjugation enhances the activity of peroxidase-mimicking DNAzymes. Bioconjug Chem 27:621–627

62. Abe H, Abe N, Shibata A, Ito K, Tanaka Y, Ito M, Saneyoshi H, Shuto S, Ito Y (2012) Structure formation and catalytic activity of DNA dissolved in organic solvents. Angew Chem Int Ed Engl 51:6475–6479

63. Li C, Zhu L, Zhu Z, Fu H, Jenkins G, Wang C, Zou Y, Lu X, Yang CJ (2012) Backbone modifications promotes peroxidase activity of G-quadruplex-based DNAzyme. Chem Commun 48:8347–8349

64. Calin GA, Sevignani C, Dumitru CD, Hyslop T, Noch E, Yendamuri S, Shimizu M, Rattan S, Bullrich F, Negrini M, Croce CM (2004) Human microRNA genes are frequently located at

fragile sites and genomic regions involved in cancers. Proc Natl Acad Sci USA 101 (9):2999–3004

65. Xiao Y, Pavlov V, Niazov T, Dishon A, Kotler M, Willner I (2004) Catalytic beacons for the detection of DNA and telomerase activity. J Am Chem Soc 126:7430–7431

66. Guo Q, Bao Y, Yang X, Wang K, Wang Q, Tan Y (2010) Amplified electrochemical DNA sensor using peroxidase-like DNAzyme. Talanta 83:500–504

67. Freeman R, Liu X, Willner I (2011) Chemiluminescent and Chemiluminescence Resonance Energy Transfer (CRET) detection of DNA, metal ions, and aptamer – substrate complexes using hemin/G-quadruplexes and CdSe/ZnS quantum dots. J Am Chem Soc 133:11597–11604

68. McKeating KS, Graham D, Faulds K (2013) Resonance Raman scattering of catalytic beacons for DNA detection. Chem Commun 49:3206–3208

69. Kosman J, Wu YT, Gluszynska A, Juskowiak B (2014) N-Methyl-4-hydrazino-7-nitrobenzofurazan: a fluorogenic substrate for peroxidase-like DNAzyme, and its potential application. Anal Bioanal Chem 406:7049–7057

70. Deng S, Cheng L, Lei J, Cheng Y, Huang Y, Ju H (2013) Label-free electrochemiluminescent detection of DNA by hybridization with a molecular beacon to form hemin/G-quadruplex architecture for signal inhibition. Nanoscale 5:5435–5441

71. Golub E, Freeman R, Niazov A, Hu J (2011) Hemin/G-quadruplexes as DNAzymes for the fluorescent detection of DNA, aptamer–thrombin complexes, and probing the activity of glucose oxidase. Analyst 136:4397–4401

72. Bi S, Jia X, Dong Y (2015) A hot-spot magnetic graphene oxide substrate for microRNA detection based on cascade chemiluminescence resonance energy transfer. Nanoscale 7:3745–3753

73. Zheng Z, Han J, Pang W, Hu J (2013) G-quadruplex DNAzyme molecular beacon for amplified colorimetric biosensing of *Pseudostellaria heterophylla*. Sensors 13:1064–1075

74. Lu N, Shao C, Deng Z (2008) Rational design of an optical adenosine sensor by conjugating a DNA aptamer with split DNAzyme halves. Chem Commun 46:6161–6163

75. Deng M, Zhang D, Zhou Y, Zhou X (2008) Highly effective colorimetric and visual detection of nucleic acids using an asymmetrically split peroxidase DNAzyme. J Am Chem Soc 130 (39):13095–13102

76. Deng M, Feng S, Luo F, Wang S, Sun X, Zhou X, Zhang XL (2012) Visual detection of *rpoB* mutations in rifampin-resistant *Mycobacterium tuberculosis* strains by use of an asymmetrically split peroxidase DNAzyme. J Clin Microbiol 50(11):3443–3450

77. Nakayama S, Sintim HO (2013) Detection of single-stranded nucleic acids via colorimetric means, using G-quadruplex probes. Methods Mol Biol 1039:153–159

78. Jiang X, Zhang H, Wu J, Yang X, Shao J, Lu Y, Qiu B, Lin Z, Chen G (2014) G-quadruplex DNA biosensor for sensitive visible detection of genetically modified food. Talanta 128:445–449

79. Fan D, Zhu J, Zhai Q, Wang E, Dong S (2016) Cascade DNA logic programmed ratiometric DNA analysis and logic devices based on a fluorescent dual-signal probe of a G-quadruplex DNAzyme. Chem Commun 52:3766–3769

80. Tian Y, Mao C (2006) Cascade signal amplification for DNA detection. Chembiochem 7:1862–1864

81. Koster DM, Haselbach D, Lehrach H, Seitz H (2011) A DNAzyme based label-free detection system for miniaturized assays. Mol BioSyst 7:2882–2889

82. Dong H, Wang C, Xiong Y, Lu H, Ju H, Zhang X (2013) Highly sensitive and selective chemiluminescent imaging for DNA detection by ligation-mediated rolling circle amplified synthesis of DNAzyme. Biosens Bioelectron 41:348–353

83. Bi S, Li L, Zhang S (2010) Triggered polycatenated DNA scaffolds for DNA sensors and aptasensors by a combination of rolling circle amplification and DNAzyme amplification. Anal Chem 82(22):9447–9454

84. Wen Y, Xu Y, Mao X, Wei Y, Song H, Chen N, Huang Q, Fan C, Li D (2012) DNAzyme-based rolling-circle amplification DNA machine for ultrasensitive analysis of microRNA in *Drosophila* larva. Anal Chem 84(18):7664–7669

85. Weizmann Y, Cheglakov Z, Willner I (2008) A Fok I/DNA machine that duplicates its analyte gene sequence. J Am Chem Soc 130:17224–17225

86. Du F, Tang Z (2011) Colrimetric detection of PCR product with DNAzymes induced by 5′-nuclease activity of DNA polymerases. Chembiochem 12:43–46

87. Li J, Yao QH, Fu HE, Zhang XL, Yang HH (2011) High sensitive and label-free colorimetric DNA detection based on nicking endonuclease-assisted activation of DNAzymes. Talanta 85:91–96

88. Fu R, Li T, Lee SS, Park HG (2011) DNAzyme molecular beacon probes for target-induced signal-amplifying colorimetric detection of nucleic acids. Anal Chem 83:494–500

89. Zhou Z, Du Y, Zhang L, Dong S (2012) A label-free, G-quadruplex DNAzyme-based fluorescent probe for signal-amplified DNA detection and turn-on assay of endonuclease. Biosens Bioelectron 34:100–105

90. Xiao HJ, Hak HC, Kong DM, Shen HX (2012) Sequence-specific detection of nucleic acids utilizing isothermal enrichment of G-quadruplex DNAzymes. Anal Chim Acta 729:67–72

91. Li H, Wu Z, Qiu L, Liu J, Wang C, Shen G, Yu R (2013) Ultrasensitive label-free amplified colorimetric detection of p53 based on G-quadruplex MBzymes. Biosens Bioelectron 50:180–185

92. Seok Y, Byun JY, Mun H, Kim MG (2014) Colorimetric detection of PCR products of DNA from pathogenic targets based on simultaneously amplified DNAzyme. Microchim Acta 181:1965–1971

93. Koo KM, Wee EJH, Rauf S, Shiddiky MJA, Trau M (2014) Microdevices for detecting locus-specific DNA methylation at CpG resolution. Biosens Bioelectron 56:278–285

94. Nie J, Zhang DW, Tie C, Zhou YL, Zhang XX (2014) G-quadruplex based two-stage isothermal exponential amplification reaction for label-free DNA colorimetric detection. Biosens Bioelectron 56:237–242

95. Chen J, Huang Y, Vdovenko M, Sakharov IY, Su G, Zhao S (2015) An enhanced chemiluminescence resonance energy transfer system based on target recycling G-quadruplex/hemin DNAzyme catalysis and its application in ultrasensitive detection of DNA. Talanta 183:59–63

96. Xu J, Qian J, Li H, Wu ZS, Shen W, Jia L (2016) Intelligent DNA machine for the ultrasensitive colorimetric detection of nucleic acids. Biosens Bioelectron 75:41–47

97. Xu Y, Li D, Cheng W, Hu R, Sang Y, Yin Y, Ding S, Ju H (2016) Chemiluminescence imaging for microRNA detection based on cascade exponential amplification machinery. Anal Chim Acta 936:229–235

98. Wu H, Liu Y, Wang H, Zhu F, Zou P (2016) Label-free and enzyme-free colorimetric detection of microRNA by catalysed hairpin assembly coupled with hybridization chain reaction. Biosens Bioelectron 81:303–308

99. Elbaz J, Moshe M, Shlyahovsky B, Willner I (2009) Cooperative multicomponent self-assembly of nucleic acid structures for the activation of DNAzyme cascades: a paradigm for DNA sensors and aptasensors. Chem Eur J 15:3411–3418

100. Liu X, Freeman R, Golub E, Willner I (2011) Chemiluminescence and Chemiluminescence Resonance Energy Transfer (CRET) aptamer sensors using catalytic hemin/G-quadruplexes. ACS Nano 5(9):7648–7655

101. Golub E, Niazov A, Freeman R, Zatsepin M, Willner I (2012) Photoelectrochemical biosensors without external irradiation: probing enzyme activities and DNA sensing using hemin/G quadruplex-stimulated Chemiluminescence Resonance Energy Transfer (CRET) generation of photocurrents. J Phys Chem C 116:13827–13834

102. Hu L, Liu X, Cecconello A, Willner I (2014) Dual switchable CRET-induced luminescence of CdSe/ZnS Quantum Dots (QDs) by the hemin/G-quadruplex-bridged aggregation and deaggregation of two-sized QDs. Nano Lett 14:6030–6035

103. Wang G, Chen L, Zhu Y, He X, Xu G, Zhang X (2014) Development of an electrochemical sensor based on the catalysis of ferrocene actuated hemin/G-quadruplex enzyme for the detection of potassium ions. Biosens Bioelectron 61:410–416

104. Xu M, Zhuang J, Chen X, Chen G, Tang G (2013) A difunctional DNA–AuNP dendrimer coupling DNAzyme with intercalators for femtomolar detection of nucleic acids. Chem Commun 49:7304–7306

105. Li T, Wang E, Dong S (2008) G-quadruplex-based DNAzyme for facile colorimetric detection of thrombin. Chem Commun 31:3654–3656

106. Li T, Wang E, Dong S (2008) Chemiluminescence thrombin aptasensor using high-activity DNAzyme as catalytic label. Chem Commun 43:5520–5522

107. Higuchi A, Siao YD, Yang ST, Hsieh PV, Fukushima H, Chang Y, Ruaan RC, Chen WY (2008) Preparation of a DNA aptamer-Pt complex and its use in the colorimetric sensing of thrombin and anti-thrombin antibodies. Anal Chem 80:6580–6586

108. Higuchi A, Yang ST, Siao YD, Hsieh PV, Fukushima H, Chang Y, Chen WY (2009) Preparation of fractioned DNA aptamer-PT complex through ultrafiltration and the colorimetric sensing of thrombin. J Membr Sci 328:97–103

109. Zhou J, Li T, Hu J, Wang E (2010) A novel dot-blot DNAzyme-linked aptamer assay for protein detection. Anal Bioanal Chem 397:2923–2927

110. Shen B, Wang Q, Zhu D, Luo J, Cheng G, He P, Fang Y (2010) G-quadruplex-based DNAzymes aptasensors for the amplified detection of thrombin. Electroanalysis 22 (24):2985–2990

111. Yuan Y, Guo X, Yuan R, Chai Y, Zhou Y, Mao L, Gan X (2011) Electrochemical aptasensor based on the dual-amplification of G-quadruplex horseradish peroxidase-mimicking DNAzyme and blocking reagent-horseradish peroxidase. Biosens Bioelectron 26:4236–4240

112. Bai L, Yuan R, Chai Y, Yuan Y, Zhuo Y, Mao L (2011) Bi-enzyme functionlized hollow PtCo nanochains as labels for an electrochemical aptasensor. Biosens Bioelectron 26:4331–4336

113. Jiang L, Yuan R, Chai Y, Yuan Y, Bai Y, Wang Y (2012) An ultrasensitive electrochemical aptasensor for thrombin based on the triplex-amplification of hemin/G-quadruplex horseradish peroxidasemimicking DNAzyme and horseradish peroxidase decorated FeTe nanorods. Analyst 138:1497–1503

114. Xie S, Chai Y, Yuan Y, Bai L, Yuan R (2014) A novel electrochemical aptasensor for highly sensitive detection of thrombin based on the autonomous assembly of hemin/G-quadruplex horseradish peroxidase-mimicking DNAzyme nanowires. Anal Chim Acta 832:51–57

115. Sun A, Qi Q, Wang X, Bie P (2014) Porous platinum nanotubes labelled with hemin/G-quadruplex based electrochemical aptasensor for sensitive thrombin analysis via the cascade signal amplification. Biosens Bioelectron 57:16–21

116. Xiao L, Chai Y, Yuan R, Wang H, Bai L (2014) Highly enhanced electrochemiluminescence based on pseudo triple-enzyme cascade catalysis and in situ generation of co-reactant for thrombin detection. Analyst 139:1030–1036

117. Padmanabhan K, Padmanabhan KP, Ferrara JD, Sadler JE, Tulinsky A (1993) The structure of alpha-thrombin inhibited by a 15-mer single-stranded DNA aptamer. J Biol Chem 268 (24):17651–17654

118. Niazov T, Pavlov V, Xiao Y, Gill R, Willner I (2004) DNAzyme-functionalized Au nanoparticles for the amplified detection of DNA or telomerase activity. Nano Lett 4 (9):1683–1687

119. Pavlov V, Xiao Y, Gill R, Dishon A, Kotler M, Willner I (2004) Amplified chemiluminescence surface detection of DNA and telomerase activity using catalytic nucleic acids labels. Anal Chem 76(7):2152–2156

120. Li T, Shi L, Wang E, Dong S (2009) Multifunctional G-quadruplex aptamers and their application to protein detection. Chem Eur J 15:1036–1042

121. Zhou WH, Zhu CL, Lu CH, Guo X, Chen F, Yang HH, Wang X (2009) Amplified detection of protein cancer biomarkers using DNAzyme functionalized nanoprobes. Chem Commun 44:6845–6847
122. Freeman R, Sharon E, Teller C, Henning A, Tzfati Y, Willner I (2010) DNAzyme-like activity of hemin–telomeric G-quadruplexes for the optical analysis of telomerase and its inhibitors. Chembiochem 11(17):2362–2367
123. Stefan L, Denat F, Monchaud D (2011) Deciphering the DNAzyme activity of multimeric quadruplexes: insights into their actual role in the telomerase activity evaluation assay. J Am Chem Soc 133(50):20405–20415
124. Wang C, Wu J, Zong C, Ju H, Yan F (2011) Highly sensitive rapid chemiluminescent immunoassay using the DNAzyme label for signal amplification. Analyst 136:4295–4300
125. Tang L, Liu Y, Ali MM, Kang DK, Zhao W, Li J (2012) Colorimetric and ultrasensitive bioassay based on a dual-amplification system using aptamer and DNAzyme. Anal Chem 84:4711–4717
126. Liu J, Lu CY, Zhou H, Xu JJ, Wang ZH, Chen HY (2013) A dual-functional electrochemical biosensor for the detection of prostate specific antigen and telomerase activity. Chem Commun 49:6602–6604
127. Jou AF, Lu CH, Ou YC, Wang SS, Hsu SL, Willner I, Ho JA (2014) Diagnosing the miR-141 prostate cancer biomarker using nucleic acid-functionalized CdSe/ZnS QDs and telomerase. Chem Sci 6:659–665
128. Zhou W, Su J, Chai Y, Yuan R, Xiang Y (2014) Naked eye detection of trace cancer biomarkers based on biobarcode and enzyme-assisted DNA recycling hybrid amplifications. Biosens Bioelectron 53:494–498
129. Hou L, Gao Z, Xu M, Cao X, Wu X, Chen G, Tang D (2014) DNAzyme-functionalized gold–palladium hybrid nanostructures for triple signal amplification of impedimetric immunosensor. Biosens Bioelectron 54:365–371
130. Li H, Fu HW, Zhao T, Kong DM (2015) Simple, PCR-free telomerase activity detection using G-quadruplex-hemin DNAzyme. RSC Adv 5:6475–6480
131. Kim NW, Piatyszek MA, Prowse KR, Harley CB, West MD, Ho PL, Coviello GM, Wright WE, Weinrich SL, Shay JW (1994) Specific association of human telomerase activity with immortal calls and cancer. Science 266:2011–2015
132. Li W, Liu Z, Lin H, Nie Z, Chen J, Xu X, Yao S (2010) Label-free colorimetric assay for methyltransferase activity based on a novel methylation-responsive DNAzyme strategy. Anal Chem 82:1935–1941
133. Zhu C, Wen Y, Peng H, Long Y, He Y, Huang Q, Li D, Fan C (2011) A methylation-stimulated DNA machine: an autonomous isothermal route to methyltransferase activity and inhibition analysis. Anal Bioanal Chem 399:3459–3464
134. Zeng Y, Hu J, Long Y, Zhang C (2013) Sensitive detection of DNA methyltransferase using hairpin probe-based primer generation rolling circle amplification-induced chemiluminescence. Anal Chem 85:6143–6150
135. Liu SC, Wu HW, Jiang JJ, Shen GL, Yu RG (2013) A novel DNAzyme-based colorimetric assay for the detection of hOGG1 activity with lambda exonuclease cleavage. Anal Methods 5:164–168
136. He K, Li W, Nie Z, Huang Y, Liu Z, Nie L, Yao S (2012) Enzyme-regulated activation of DNAzyme: a novel strategy for a label-free colorimetric DNA ligase assay and ligase-based biosensing. Chem Eur J 18:3992–3999
137. Jiang HX, Kong DM, Shen HX (2014) Amplified detection of DNA ligase and polynucleotide kinase/phosphatase on the basis of enrichment of catalytic G-quadruplex DNAzyme by rolling circle amplification. Biosens Bioelectron 55:133–138
138. Zhou Z, Peng L, Wang X, Xiang Y, Tong A (2014) A new colorimetric strategy for monitoring caspase 3 activity by HRP-mimicking DNAzyme–peptide conjugates. Analyst 139:1178–1183

139. Liu Z, Li W, Nie Z, Peng F, Huang Y, Yao S (2014) Randomly arrayed G-quadruplexes for label-free and real-time assay of enzyme activity. Chem Commun 50:6875–6878
140. Jiang C, Yan C, Jiang J, Yu R (2013) Colorimetric assay for T4 polynucleotide kinase activity based on the horseradish peroxidase-mimicking DNAzyme combined with λ exonuclease cleavage. Anal Chim Acta 766:88–93
141. Shi B, Qin Y, Huang M, Zhao J, Su Y, Zhao S (2015) A G-quadruplex-based colorimetric assay of S1 nuclease activity and inhibition. Anal Methods 7:5600–5605
142. Li T, Wang E, Dong S (2009) G-quadruplex-based DNAzyme as a sensing platform for ultrasensitive colorimetric potassium detection. Chem Commun 5:580–582
143. Fan X, Li H, Zhao J, Lin F, Zhang L, Zhang Y, Yao S (2012) A novel label-free fluorescent sensor for the detection of potassium ion based on DNAzyme. Talanta 89:57–62
144. Wang H, Wang DM, Gao MX, Wang J, Huang CZ (2014) Potassium-induced G-quadruplex DNAzyme as a chemiluminescent sensing platform for highly selective detection of K$^+$. Anal Methods 6:7415–7419
145. Li T, Dong S, Wang E (2009) Label-free colorimetric detection of aqueous mercury ion (Hg^{2+}) using Hg^{2+}-modulated G-quadruplex-based DNAzymes. Anal Chem 81:2144–2149
146. Hao Y, Guo Q, Wu H, Guo L, Zhong L, Wang J, Lin T, Fu F, Chen G (2014) Amplified colorimetric detection of mercuric ions through autonomous assembly of G-quadruplex DNAzyme nanowires. Biosens Bioelectron 52:261–264
147. Tang X, Wang YS, Xue JH, Zhou B, Cao JX, Chen SH, Li MH, Wang XF, Zhu YF, Huang YQ (2015) A novel strategy for dual-channel detection of metallothioneins and mercury based on the conformational switching of functional chimera aptamer. J Pharm Biomed Anal 107:258–264
148. Zhou XH, Kong DM, Shen HX (2010) G-quadruplex-hemin DNAzyme-amplified colorimetric detection of Ag$^+$ ion. Anal Chim Acta 678:124–127
149. Zhou XH, Kong DM, Shen HX (2010) Ag$^+$ and cysteine quantitation based on G-quadruplex-hemin disruption by Ag$^+$. Anal Chem 82:789–793
150. Zhang K, Wang K, Zhu X, Xie M (2015) Sensitive and selective amplified detection of silver ion based on NEase-aided target recycling. RSC Adv 5:89047–89051
151. Liu G, Yuan Y, Wang J (2016) Hemin/G-quadruplex DNAzyme nanowires amplified luminol electrochemiluminescence system and its application in sensing silver ions. RSC Adv 6:37221–37225
152. Elbaz J, Shlyahovsky B, Willner I (2008) A DNAzyme cascade for the amplified detection of Pb^{2+} ions or L-histidine. Chem Commun 13:1569–1571
153. Wang Y, Wang J, Yang F, Yang X (2010) Spectrophotometric detection of lead(II) ion using unimolecular peroxidase-like deoxyribozyme. Microchim Acta 171:195–201
154. Li F, Yang L, Chen M, Qian Y, Tang B (2013) A novel and versatile sensing platform based on HRP-mimicking DNAzyme-catalyzed template-guided deposition of polyaniline. Biosens Bioelectron 41:903–906
155. Wang H, Wang DM, Huang CZ (2015) Highly sensitive chemiluminescent detection of lead ion based on its displacement of potassium in G-quadruplex DNAzyme. Analyst 140:5742–5747
156. Zhu X, Gao X, Liu Q, Lin Z, Qiu B, Chen G (2011) Pb^{2+}-introduced activation of horseradish peroxidase (HRP)-mimicking DNAzyme. Chem Commun 47:7437–7439
157. Zhou Q, Lin Y, Wei Q, Chen G, Tang D (2016) Highly sensitive electrochemical sensing platform for lead ion based on synergetic catalysis of DNAzyme and Au–Pd porous bimetallic nanostructures. Biosens Bioelectron 78:236–243
158. Xue S, Jing P, Xu W (2016) Hemin on graphene nanosheets functionalized with flower-like MnO$_2$ and hollow AuPd for the electrochemical sensing lead ion based on the specific DNAzyme. Biosens Bioelectron 86:958–965
159. Teller C, Shimron S, Willner I (2009) Aptamer-DNAzyme hairpins for amplified biosensing. Anal Chem 81:9114–9119

160. Wang G, Chen L, Zhu Y, He X, Xu G, Zhang X (2014) Prussian blue-Au nanocomposites actuated hemin/G-quadruplexes catalysis for amplified detection of DNA, Hg^{2+} and adenosine triphosphate. Analyst 139:5297–5303

161. Wang G, Chen L, Zhu Y, Wang L, Zhang X (2014) Adenosine triphosphate sensing by electrocatalysis with DNAzyme. Electroanalysis 26:312–318

162. Chu Z, Zhang L, Huang Y, Zhao S (2014) A G-quadruplex DNAzyme chemiluminescence aptasensor based on the target triggered DNA recycling for sensitive detection of adenosine. Anal Methods 6:3700–3705

163. Wu Q, Shen H, Sun Y, Song L (2016) Study on sensing strategy and performance of a microfluidic chemiluminescence aptazyme sensor. Talanta 150:531–538

164. Yang C, Lates V, Prieto-Simon B, Marty JL, Yang X (2012) Aptamer-DNAzyme hairpins for biosensing of Ochratoxin A. Biosens Bioelectron 32:208–212

165. Yang C, Lates V, Prieto-Simon B, Marty JL, Yang X (2013) Rapid high-throughput analysis of ochratoxin A by the self-assembly of DNAzyme–aptamer conjugates in wine. Talanta 116:520–526

166. Shim WB, Mun H, Joung HA, Ofori JA, Chung DW, Kim MG (2014) Chemiluminescence competitive aptamer assay for the detection of aflatoxin B1 in corn samples. Food Control 36:30–35

167. Mun H, Jo EJ, Li T, Joung HA, Hong DG, Shim WB, Jung C, Kim MG (2014) Homogeneous assay of target molecules based on chemiluminescence resonance energy transfer (CRET) using DNAzyme-linked aptamers. Biosens Bioelectron 58:308–313

168. Wang C, Dong X, Liu Q, Wang K (2015) Label-free colorimetric aptasensor for sensitive detection of ochratoxin A utilizing hybridization chain reaction. Anal Chim Acta 860:83–88

169. Seok Y, Byun JY, Shim WB, Kim MG (2015) A structure-switchable aptasensor for aflatoxin B1 detection based on assembly of an aptamer/split DNAzyme. Anal Chim Acta 886:182–187

170. Jo EJ, Mun H, Kim SJ, Shim WB, Kim MG (2015) Detection of ochratoxin A (OTA) in coffee using chemiluminescence resonance energy transfer (CRET) aptasensor. Food Chem 194:1102–1107

171. Zhang DW, Nie J, Zhang FT, Xu L, Zhou YL, Zhang XX (2013) Novel homogeneous label-free electrochemical aptasensor based on functional DNA hairpin for target detection. Anal Chem 85:9378–9382

172. Nie J, Zhang DW, Tie C, Zhou YL, Zhang XX (2014) A label-free DNA hairpin biosensor for colorimetric detection of target with suitable functional DNA partners. Biosens Bioelectron 49:236–242

173. Hou T, Li C, Wang X, Zhao C, Li F (2013) Label-free colorimetric detection of coralyne utilizing peroxidase-like split G-quadruplex DNAzyme. Anal Methods 5:4671–4674

174. Hou T, Wang X, Liu S, Du Z, Li F (2013) A label-free and colorimetric turn-on assay for coralyne based on coralyne-induced formation of peroxidase-mimicking split DNAzyme. Analyst 138:4728–4731

175. Zhang H, Jiang B, Xiang Y, Chai Y, Yuan R (2012) Label-free and amplified electrochemical detection of cytokine based on hairpin aptamer and catalytic DNAzyme. Analyst 137:1020–1023

176. Zhou W, Gong X, Xiang Y, Yuan R, Chai Y (2013) Target-triggered quadratic amplification for label-free and sensitive visual detection of cytokines based on hairpin aptamer DNAzyme probes. Anal Chem 86:953–958

177. Omar N, Loh Q, Tye GJ, Choong YS, Noordin R, Gloker J, Lim TS (2014) Development of an antigen-DNAzyme based probe for a direct antibody-antigen assay using the intrinsic DNAzyme activity of a daunomycin aptamer. Sensors 14:346–355

178. Bo H, Wang C, Gao Q, Zhang C (2013) Selective, colorimetric assay of glucose in urine using G-quadruplex-based DNAzymes and 10-acetyl-3,7-dihydroxy phenoxazine. Talanta 108:131–135

179. Hu Y, Wang F, Lu CH, Girsh J, Golub E, Willner I (2014) Switchable enzyme/DNAzyme cascades by the reconfiguration of DNA nanostructures. Chem Eur J 20:16203–16209
180. Yang DK, Kou CJ, Chen LC (2015) Synthetic multivalent DNAzymes for enhanced hydrogen peroxide catalysis and sensitive colorimetric glucose detection. Anal Chim Acta 856:96–102
181. Li R, Xiong C, Xiao Z, Ling L (2012) Colorimetric detection of cholesterol with G-quadruplex-based DNAzymes and ABTS^{2-}. Anal Chim Acta 724:80–85
182. Wang M, Han Y, Nie Z, Lei C, Huang Y, Gou M, Yao S (2010) Development of a novel antioxidant assay technique based on G-quadruplex DNAzyme. Biosens Bioelectron 26:523–529

Adv Biochem Eng Biotechnol (2020) 170: 85–106
DOI: 10.1007/10_2017_37
© Springer International Publishing AG 2017
Published online: 16 November 2017

Hemin/G-Quadruplex Horseradish Peroxidase-Mimicking DNAzyme: Principle and Biosensing Application

Negar Alizadeh, Abdollah Salimi, and Rahman Hallaj

Abstract Enzymes are macromolecular biological catalysts that accelerate chemical reactions. Enzyme labels are commonly used to obtain signal amplification in sensors and biosensors on the basis of reactions of some enzymes such as horseradish peroxidase (HRP). However, use of natural enzymes can encounter some challenges. Lately, nucleic acids that exhibit catalytic properties have attracted growing interest because they have certain advantages in comparison with traditional protein enzymes. DNAzymes are DNA-based catalysts, representing an important class of functional DNA, which have been widely used because of their excellent activity, programmability, signal amplification through catalytic turnover, high chemical stability, simple synthesis, and easy modification. Considering these remarkable properties, the hemin/G-quadruplex DNAzyme is extensively used in electrochemical, colorimetric, and chemiluminescence sensors and biosensors for detection of various targets.

Keywords DNAzymes, Enzyme, Hemin/G-quadruplex, HRP, Sensor

N. Alizadeh
Department of Chemistry, University of Kurdistan, 66177-15175 Sanandaj, Iran

A. Salimi (✉) and R. Hallaj
Department of Chemistry, University of Kurdistan, 66177-15175 Sanandaj, Iran

Research Center for Nanotechnology, University of Kurdistan, 66177-15175 Sanandaj, Iran
e-mail: absalimi@yahoo.com

Contents

1 Introduction

One of the most important molecules in the animated world is DNA. DNA molecules have met various applications in addition to its primary role as genetic storage. Throughout history, DNA replication and transcription mechanisms have been studied, and yet relatively little is known about this molecule. Almost 30 years ago, nucleic acids were used in materials science, which opened the door to applications in various fields of science [1, 2]. The chemical structure of DNA is shown in Fig. 1.

Other than the typical Watson–Crick base pair, nucleotides can form different higher-order structures of DNA molecules, which can be employed in bioanalytical applications [3].

Ribozymes (catalytic RNAs) are catalyzed in many cell reactions, such as the hydrolysis of phosphodiester bonds or transamination, but DNA does not exhibit such activity, likely because of differences in structural flexibility. The $2'$-hydroxyl group of ribose, which is the major determinant of stable duplex formation, structural flexibility, and the functions of RNA does not exist in the DNA molecule. Nevertheless, it was assumed that single-stranded DNA sequences could also show enzymatic activity, and the first deoxyribozyme was synthesized in 1994, presented by Breaker and Joyce [4]. All deoxyribozymes are artificially obtained and are not found in vivo. Systematic evolution of ligands by exponential enrichment (SELEX) is a procedure of in vitro selection, providing a tool for developing oligonucleotides that can catalyze specific reactions. However, there are numerous natural protein enzymes, and researchers look for other catalytic molecules, because of several advantages of deoxyribozymes over classical enzymes. First, DNA exhibits catalytic activity in a wide temperature range and preserves its activity even at high temperatures, whereas enzymes are susceptible to temperature variations. Second, proteins require complicated preparation and purification, unlike oligonucleotides, which can be easily obtained using chemical synthesis and the PCR technique [5]. In addition, DNAzyme shows high specificity for its substrate strand and even one single base mismatch in the antisense arms significantly decreases the cleaving activity.

One kind of DNAzyme is named G-quadruplex-DNAzyme. In this DNAzyme, a guanine (G)-rich nucleic acid sequence in particular is predisposed to forming higher order structures and folds into a parallel or an antiparallel G-quadruplex in the presence of K^+, Pb^{2+}, or NH_4^+. Recently, it has been demonstrated that G-rich

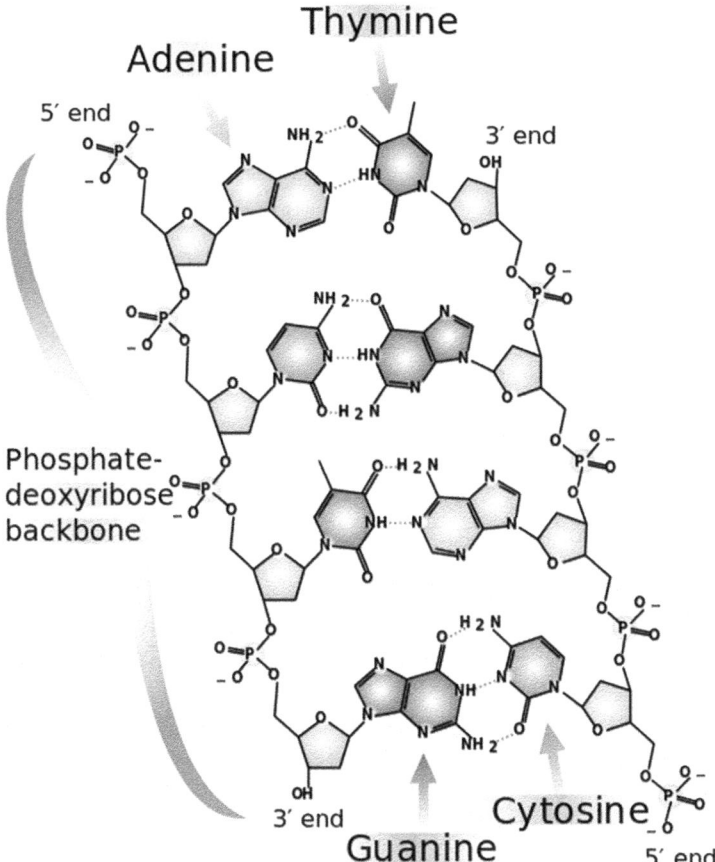

Fig. 1 Chemical structure of DNA

sequence could bind to hemin to form the hemin/G-quadruplex HRP-mimicking DNAzyme, which can catalyze the reduction of H_2O_2. Therefore, hemin/G-quadruplex DNAzyme has been extensively used as a catalytic label in place of HPR, as the amplified detection and the coupling of DNAzymes with nanomaterials produces new approaches in the designing of biosensors or nanodevices [6, 7].

2 Structure and Activity of Peroxidase-Mimicking DNAzymes

Sen and co-workers [8] first reported DNAzymes with peroxidase-like activity in the late 1990s. They considered that enzymes containing hemin as a cofactor showed peroxidase activity and that hemin singly, even without apoenzyme,

Table 1 Examples of DNA aptamers that form G-quadruplexes and enhance peroxidase activity of hemin

DNA aptamer	Sequence
PS5.M	5′-GTGGGGCATTGTGGGTGGGTGTGG-3′
PS2.M	5′-GTGGGTAGGGCGGGTTGG-3′
T4G4	5′-TTTTGGGG-3′
Hum21	5′-GGGTTAGGGTTAGGGTTAGGG-3′
AGRO 100	5′-GGTGGTGGTGGTTGTGGTGGTGGTGG-3′

could catalyze peroxidase reactions. The search began for an artificial system that would increase the catalytic activity of hemin. This research led to DNA oligomers being identified as appropriate molecules that could form stable complexes with hemin, and these systems appeared to enhance the catalytic activity of hemin. The oligonucleotide DNA contains a guanine-rich sequence and forms a quadruplex structure with hydrogen bonds between the guanine bases. It is well known that typical hydrogen bonds are formed between bases in nucleic acids. This is the most common type of pairing, called Watson–Crick base pair. However, nucleic acids can also take part in another kind of pairing known as Hoogsteen base pairing. This kind of hydrogen bonding terminates in the formation of higher-order structures similar to the quadruplex structures mentioned previously. Some of the first quadruplex-forming sequences that could enhance the peroxidase activity of hemin are reported in Table 1 [9–11].

Quadruplexes can show many different topologies. These structures can be formed by one, two or four DNA molecules and are very stable, which makes them useful in biotechnology applications. Some examples of G-quadruplex structures are presented in Fig. 2 [12].

Shangguan and co-workers [13] discovered the relationship between G-quadruplex conformation and the catalytic activity of DNAzymes, which was related to the ability to bind hemin. Tetramolecular structures with parallel strands show low peroxidase action, and unimolecular antiparallel quadruplexes exhibit low enzymatic activity. The highest activity was observed for parallel or mixed intramolecular quadruplexes [13, 14]. Hemin binds with external guanines in a quadruplex, so that parallel structure is more favorable for the binding of ligands through end-stacking. Lower activity of antiparallel quadruplexes is likely related to a steric barrier caused by loops, which makes hemin binding difficult [15]. The topology of G-quadruplexes is also affected by factors such as the type of coordinated cation, the number of G-tetrads, and the loop composition. The presence of cations is necessary for quadruplex formation. Early studies indicated that the potassium ion was essential for quadruplex formation and catalytic activity [16]. However, recent reports revealed that other ions could also be used and in some instances replace K^+ with NH_4^+ ions positively influenced by enzymatic activity. Nakayama and Sintim [17] also noticed that the presence of $NH4^+$ reduced the danger of oxidative degradation of oligonucleotides. The sequence of loops in a quadruplex structure could also impress the stability, conformation, and

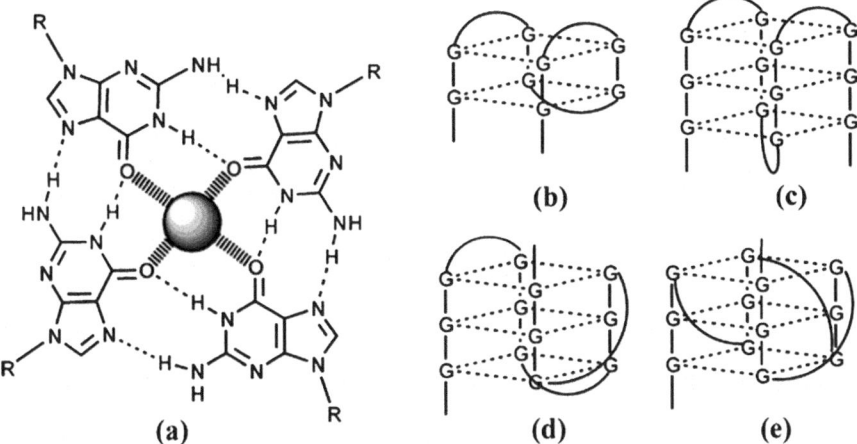

Fig. 2 (**a**) Structure of G-tetrad showing hydrogen bonds between four guanines and the interactions with a cation and (**b**) schematic representation of G-quadruplex structures: an antiparallel "chair-type" G-quadruplex with all lateral loops, (**c**) an antiparallel "basket-type" G-quadruplex with one diagonal and two lateral loops, (**d**) a hybrid-type quadruplex with parallel–antiparallel loops orientation, and (**e**) a parallel quadruplex with all loops positioned alongside the grooves (from Juskowiak [12])

activity of DNAzymes. It has been found that oligonucleotides with longer loops generate antiparallel quadruplexes and exhibit lower activity [17].

Performing an appropriate indicator reaction is needed to observe the catalytic activity of DNAzymes. In biological applications using horseradish peroxidase, TMB (3,3,5,5-tetramethylbenzidine sulfate) and luminol are commonly utilized as substrates [18, 19]. The oxidation of TMB in the presence of H_2O_2 produces a colored product, and changes in the absorption spectrum followed to monitor the reaction progress. The reaction of luminol with hydrogen peroxide generates chemiluminescence (CL), for which the intensity serves as an analytical signal (Fig. 3).

3 A Review of the Construction of G-Quadruplex-Based Sensing Platforms

3.1 Electrochemical Sensors and Biosensors

Electrochemical sensors are used in one of the major analytical methods for sensitive and selective detection of proteins, biomolecules such as DNA, and many other molecules and ions as targets. Nowadays, the increasing requirement for detection of ultralow amounts of analytes is pushing the enhancement of sensitivity by selecting different signal amplification strategies [20, 21]. Enzyme-based sensors with excellent

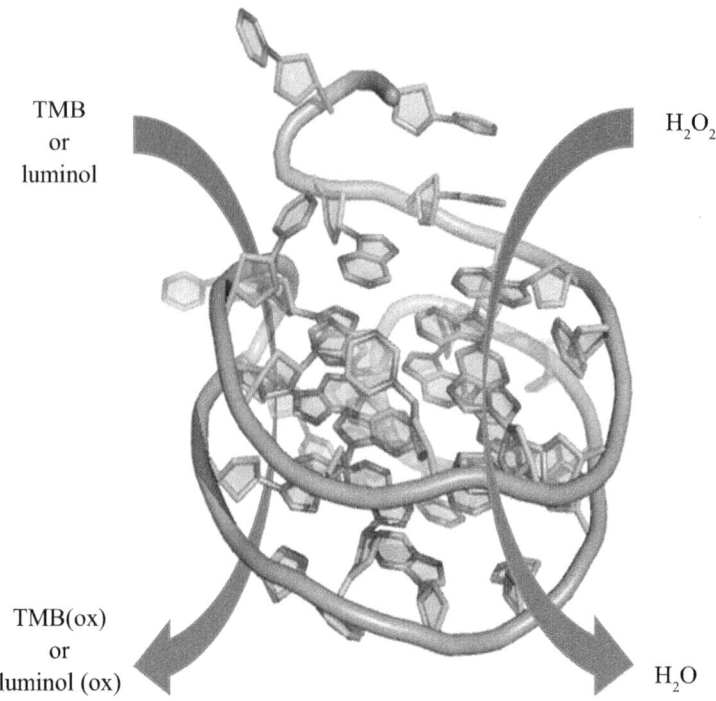

Fig. 3 Scheme of indicator reactions catalyzed by peroxidase-mimicking DNAzymes: the oxidation of TMB and luminol (from Kong et al. [18])

properties such as stability, reusability, and availability for almost any given analyte target have attracted the substantial regard of researchers and emerged as viable labels for detecting and quantifying trace amounts of analytes [22–24]. Horseradish peroxidase or HRP, glucose oxidase or GOx and alkaline phosphatase or ALP are the most commonly used enzymes label in an electrochemical study of the biocatalytic reaction [25]. However, using the natural enzyme can produce a nonconductive insoluble product on the electrode surface, and the insoluble precipitation can block the electron transfer process of redox probes, in addition, the assembling of these enzymes on the nanocarriers has become a bottle-neck because of their limited stability. Artificial mimicking enzymes, especially DNAzymes, have been used as catalytic amplifiers in biosensing strategies. Similar to natural enzymes, DNAzymes show high catalytic activities toward specific substrates whereas they are more stable than natural enzymes and can be denatured and renatured several times without losing their catalytic activity [26, 27]. Based on this merit Li et al. reported construction of electrochemical DNA biosensors based on HRP-mimicking hemin/G-quadruplex wrapped GOx nanocomposites as tags for the detection of *Escherichia coli* O157:H7 [28]. The procedure for the proposed DNA biosensor is shown in Fig. 4.

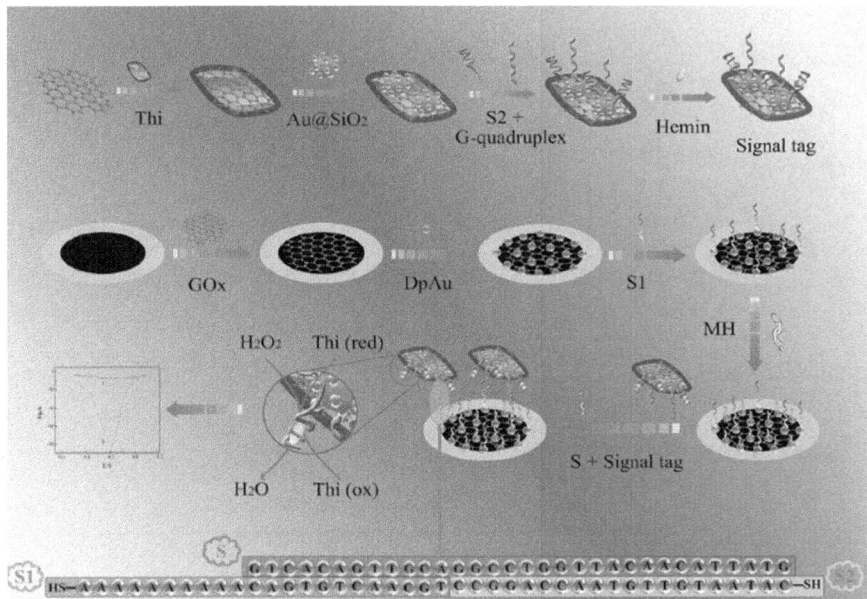

Fig. 4 Schematic diagram of the preparation of GOx-Thi-Au@SiO$_2$ nanocomposites and the fabrication of the DNA biosensor (from Li et al. [28])

In this work, the hemin/G-quadruplexes and S$_2$ were united on the GOx-Thi-Au@SiO$_2$-formed nanocomposites to achieve the signal amplification and the sensitive detection was monitored by a differential pulse voltammetry (DPV) signal. To prepare this DNA biosensor, the cleaned GCE electrode was coated with GOx; to attach primary S$_1$, a nano-Au layer was formed by electrodeposition in HAuCl$_4$ solution. Therefore, these modified electrodes were immersed in S$_1$; to eliminate the nonspecific binding effects of the electrode surface, the S$_1$/nano-Au/GOx/GCE electrode was dropped on (6-Mercapto-1-hexanol) MH solution. The as-prepared DNA biosensor, MH/S1/nano-Au/GOx/GCE, was incubated with *E. coli* O157:H7 samples and S$_2$/GOx-Thi-Au@SiO2/DNAzyme solution respectively. Ultimately, the electrochemical detection was carried out in an electrolytic cell containing PBS (pH = 7.4) and H$_2$O$_2$.

Alizadeh et al. also developed a sensitive electrochemical immunosensor for hepatitis B virus surface antigen detection based on hemin/G-quadruplex horseradish peroxidase-mimicking DNAzyme signal amplification [29].

The sensing mechanism of the proposed immunosensor was shown in Fig. 5. Fe$_3$O$_4$ was used to immobilize the primary antibody (Ab$_1$); thus, the Fe$_3$O$_4$-Ab$_1$ was captured on the glassy carbon electrode (GCE) surface by a magnetic underside. Then, the GCE/Fe$_3$O$_4$-Ab$_1$ electrode was incubated with different solutions of HBsAg and subsequently with Ab$_2$-MB-Au-DNAzyme as a signal amplifier. G-quadruplex DNAzyme can catalyze the reduction of H$_2$O$_2$ with the help of methylene blue (MB), so that the reduction peak current of H$_2$O$_2$ increased with

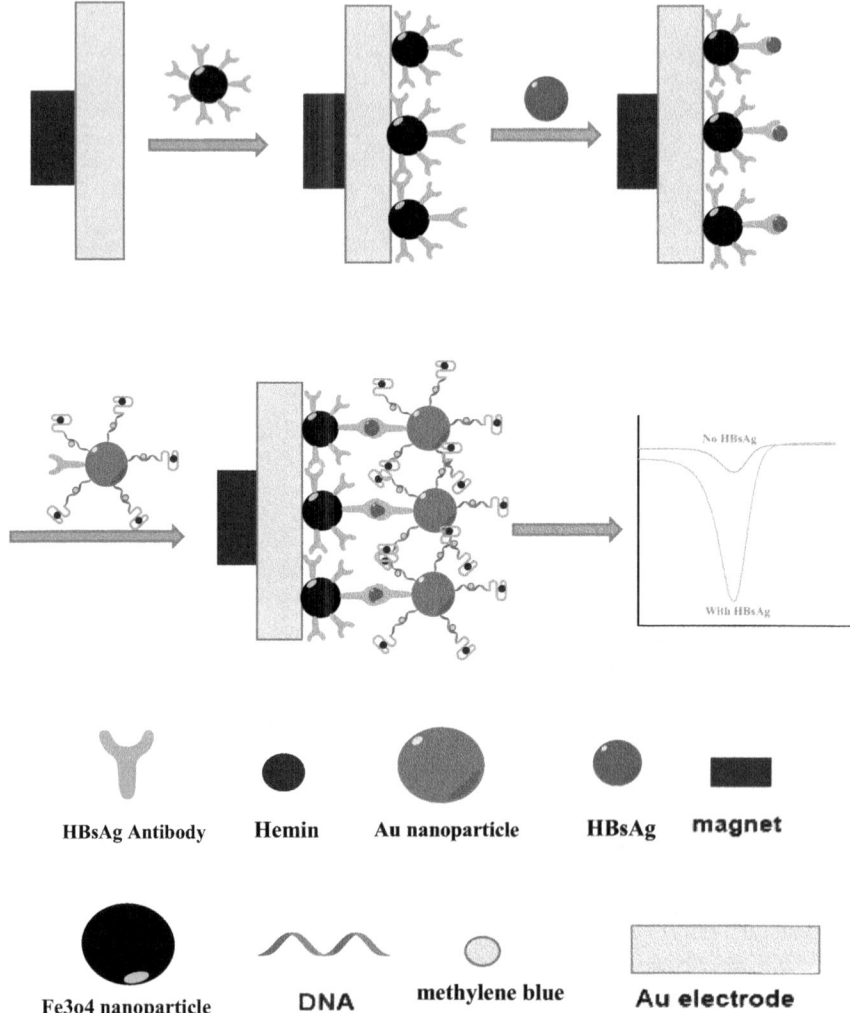

Fig. 5 Different steps for the fabrication of HBsAg immunosensor (from Alizadeh et al. [29])

increasing HBsAg concentration. The immunosensor response is based on the following redox process.

$$DNAzyme(red) + H_2O_2 \rightarrow DNAzyme(ox) + H_2O \tag{1}$$

$$Methylene\ blue(red) + DNAzyme(ox)$$
$$\rightarrow DNAzyme\ (red) + methylene\ blue\ (ox) \tag{2}$$

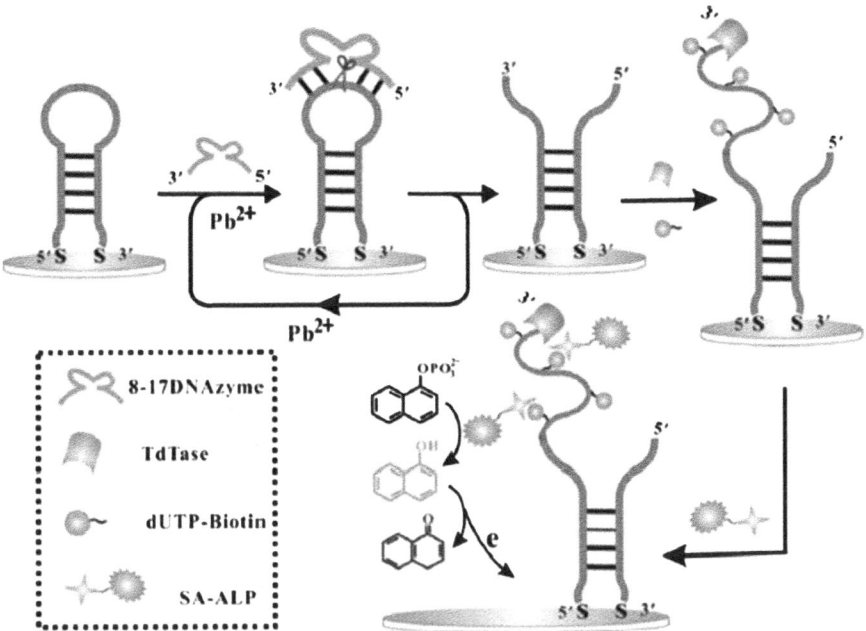

Fig. 6 Schematic illustration of the electrochemical DNAzyme sensor for the lead ion based on the TdTase-mediated extension and alkaline phosphatase amplification (from Liu et al. [33])

$$\text{Methylene blue } (\text{ox}) + 2e- \rightarrow \text{methylene blue (red)} \qquad (3)$$

It has been emerged for the DNAzyme sensor fabrication toward different metal ions for example Mg^{2+}, Na^+, Pb^{2+}, UO_2^{2+}, and lanthanides, owing to the specific requirement of metal cofactors for RNA-cleaving DNAzyme during the in vitro selection process [30]. The 8-17 DNAzyme, one of the RNA-cleaving DNAzymes, is a Pb^{2+}-dependent DNAzyme, and has been implemented to develop a different type of sensors for Pb^{2+} detection [31, 32].

Liu et al. reported an electrochemical DNAzyme sensor for the lead ion based on cleavage-induced, template-independent polymerization and alkaline phosphatase amplification [33]. As described in Fig. 6, The hairpin-like substrate DNA strand of the 8-17 DNAzyme was first immobilized on the electrode and hitherto no exposed 3′-OH could be used as the primer of terminal deoxynucleotidyl transferase (TdTase). In the presence of Pb^{2+} and the catalytic strand of 8-17 DNAzyme, the HP DNA could be cleaved and the 3′-OH terminal could be created and then operated by TdTase for the base extension to combine biotinylated dUTP (dUTP-biotin). The streptavidin-labeled alkaline phosphatase (SA-ALP) could be further incorporated with the extended biotin. The applied ALP then catalyzes conversion of electrochemically inactive 1-naphthyl phosphate (1-NP) into an electrochemically active phenol to produce an amplified electrochemical signal.

It has been reported that the hemin/G-quadruplex DNAzyme can simultaneously serve as an NADH oxidase and HRP-mimicking DNAzyme [34]. Based on this observation, Yuan and co-workers proposed a sensitive pseudobienzyme electrocatalytic DNA biosensor for Hg^{2+} detection by using the autonomously assembled hemin/G quadruplex DNAzyme nanowires for signal amplification [35]. As displayed in Fig. 7, thiol-functionalized capture DNA was firstly immobilized on a nano-Au-modified glass carbon electrode (GCE). Owing to the coordination chemistry for the T-Hg^{2+}-T complexes, in the presence of Hg^{2+} the primer DNA could be assembled on the electrode, which results in triggering of the HCR to form the hemin/G-quadruplex DNAzyme nanowires with the substantial redox probe thionine (Thi). At first, the hemin/G-quadruplex nanowires acted as an NADH oxidase to assist the concomitant formation of H_2O_2 in the presence of dissolved O_2. Next, with the redox probe Thi as electron mediator, the hemin/G-quadruplex nanowires acted as an HRP-mimicking DNAzyme that quickly bioelectrocatalyzed the reduction of the produced H_2O_2, which led to a significantly amplified electrochemical signal. The designed biosensor could be also extended toward the on-site monitoring of the other metal ions or trace pollutants in environmental matrices by using different DNA or aptamer probes.

Adenosines are endogenous nucleosides that play a fundamental role in biochemical processes and in signal transduction [36]; thus, it is an agent for establishing a method of detecting adenosines in biochemical studies. Wang et al. constructed a dual-signal and dual-functional electrochemical sensor for ATP and H_2O_2 [37]. The detailed principle of ATP and H_2O_2 detection is shown in Fig. 8.

At first, the anti-ATP aptamer conformation changed after its binding to ATP, then the anti-ATP aptamer formed a stable G-quadruplex structure after the addition of ATP owing to its guanine-rich nucleic acid sequence. Second, with the addition of hemin, a peroxidase-like G-quadruplex configuration was formed as a DNAzyme. The anti-ATP aptamer was immobilized on gold electrodes through thiol-gold chemistry; only a weak signal appeared because of the long distance between the Fc molecule and the electrode for electron exchange. Reaction with ATP made the attached Fc approach the electrode and an electrochemical signal could be observed corresponding to the redox reaction of Fc. To solve the problem of unsatisfactory sensitivity in the electrochemical switchable ATP sensing, the hemin/G-quadruplex that was formed was used for reduction of H_2O_2 and amplification of the current.

With hemin/G-quadruplex DNAzyme as a label many signal amplification approaches have been reported. For another example, Meng's group developed an electrochemical biosensor of microRNA-21 based on the bio bar code and hemin/G-quadruplet DNAzyme [38]. In this research, MiRNA-21 could be determined indirectly by the reduction response of benzoquinone (Fig. 9). At the first part, AuNPs were deposited in HAuCl4 solution containing KNO_3 by amperometry; the electrode obtained was named AuNPs/Au. Then, the probe DNA and blocking reagent were immobilized on the surface via bonding of Au-S. After the hybridization between miRNA-21 and hairpin DNA, the hemin/G-quadruplex structure was formed, resulting in an improved signal owing to its excellent catalytic ability.

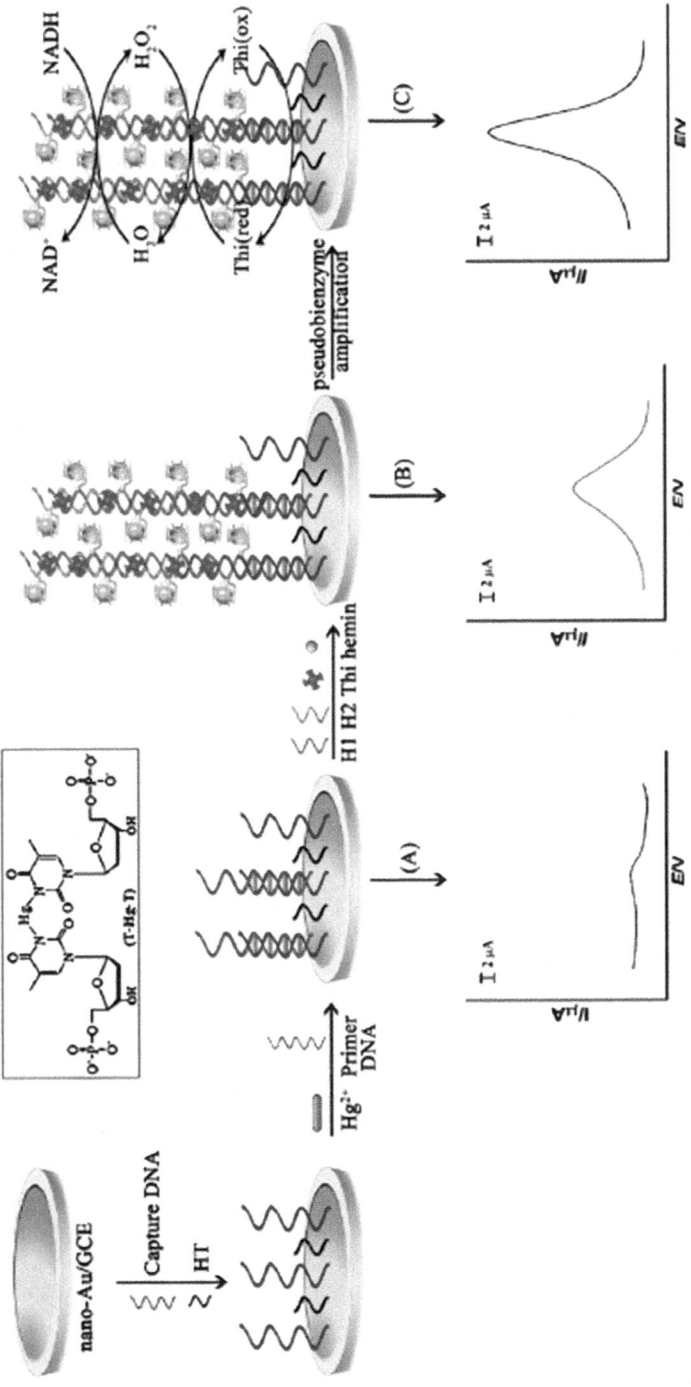

Fig. 7 Schematic illustration of the sensitive pseudobienzyme electrocatalytic DNA biosensor for Hg^{2+} detection (from Yuana et al. [35])

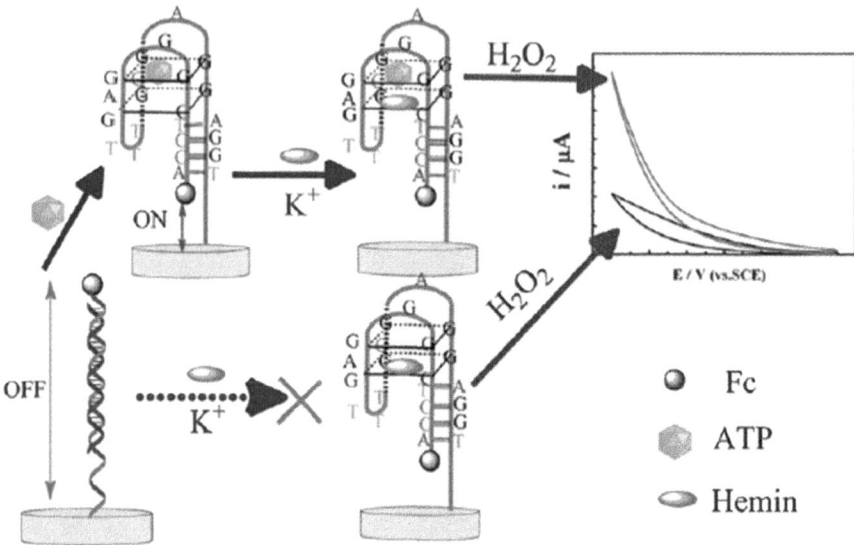

Fig. 8 The assay principle of the dual-functional electrochemical biosensor developed for the detection of ATP and H_2O_2 (from Wang et al. [37])

3.2 Colorimetric Sensors and Biosensors

In recent years, colorimetric biosensing has attracted much attention because of its intrinsic advantages such as low cost, simplicity, practicality, its potential application in point-of-care diagnosis, and even only utilizing the naked eye [39, 40].

From the view of the privilege of colorimetric sensor, Li et al. constructed a G-quadruplex DNAzymes-based platform for selective and sensitive colorimetric sensing of free heme in rat brain [41].

PS2.M (5′-GTGGGTAGGGCGGGTTGG-30), an 18-base G-rich DNA sequence in the presence of K^+, can form DNAzyme with high peroxidase-like activity; K^+-stabilized PS2.M was utilized to sense free heme in the cerebral system (Fig. 10). By the addition of K^+, the conformation of PS2.M changes from a random coil to a "parallel" G-quadruplex structure, which can bind free heme in the cerebral system with high affinity. The resulting heme/G-quadruplex complex exhibits high peroxidase-like activity, which can catalyze the oxidation of colorless $ABTS^{2-}$ to green $ABTS^{\bullet-}$ in the presence of H_2O_2. The presence of free heme in the cerebral system is therefore detected by the absorption enhancement of green $ABTS^{\bullet-}$.

Based on the DNAzyme probe self-assembled gold nanoparticles, Luo's group developed a colorimetric assay method for the *invA* gene of *Salmonella* [42]. The principle of *Salmonella* detection was illustrated in Fig. 11. The plate is coated with a layer of reactive N-oxysuccinimide ester (NOS groups) in a slightly alkaline environment; afterward, the capture probe that modified with a primary amino group was covalently linked to the plate. In the presence of the target sequence, the hybridization

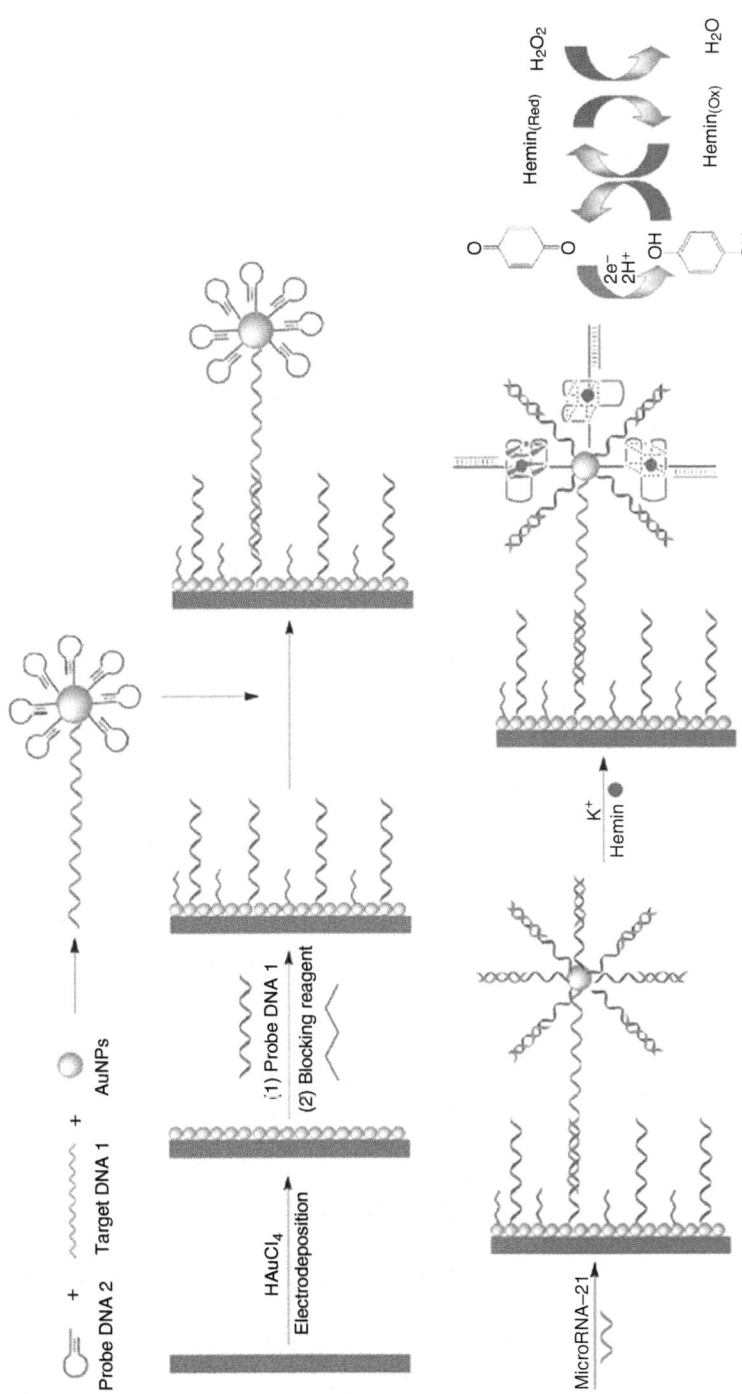

Fig. 9 Schematic representation of miRNA-21 biosensor fabrication (from Meng et al. [38])

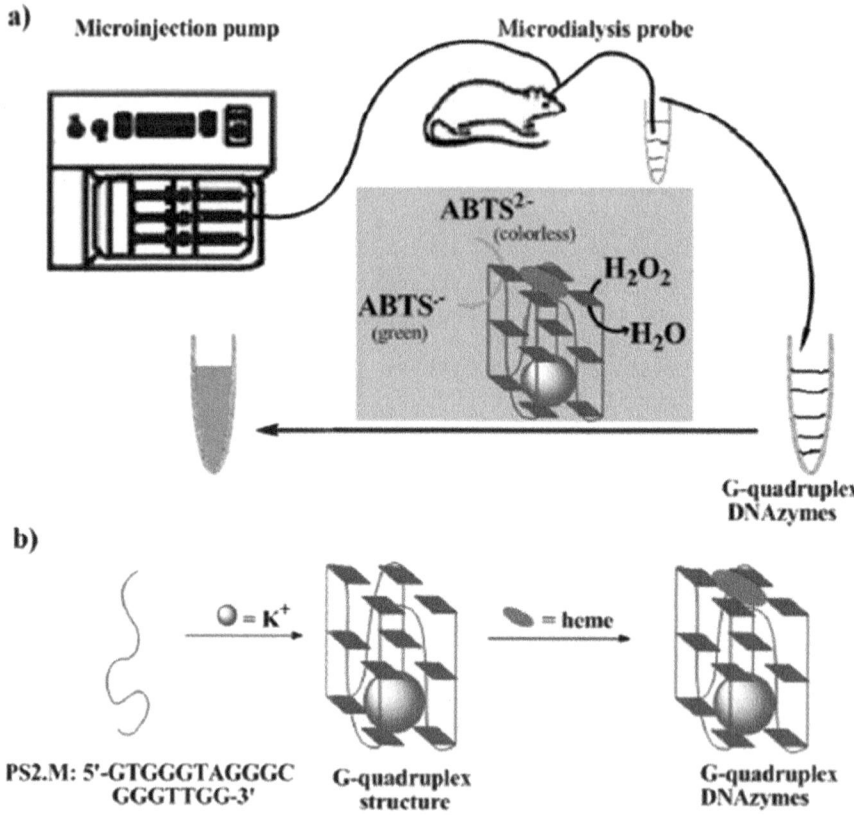

Fig. 10 G-quadruplex DNAzymes-based colorimetric sensing of cerebral free heme coupled with in vivo microdialysis (from Li et al. [41])

reaction between the capture probe and the detection probefunctionalized AuNPs occurred and a sandwich-type hybridization structure formed. Then, by the addition of hemin, the DNAzyme probe assembled on AuNPs could form a G-quadruplex/hemin complex that can catalyze the H_2O_2-mediated oxidation of 3,3',5,5'-tetramethylbenzidine (TMB) to cause an exhibitive color change.

Up to now, G-quadruplex-DNAzyme labels have been used extensively for the detection of a number of analytes, including small molecules [19], protein [43], DNA [44], and metal ion [45]. Owing to the ever-increasing environmental and health worries about Hg^{2+}, it is an agent for developing a monitoring routine for trace Hg^{2+} detection with high sensitivity, selectivity, cost efficiency, and straightforward operation. Ren et al. reported a label-free colorimetric assay for the sensitive detection of Hg^{2+} based on Hg^{2+}-triggered hairpin DNA probe (H-DNA) termini-binding and exonuclease III (Exo III)-assisted target recycling, in addition to hemin/G-quadruplex (DNAzyme) signal amplification [46]. Design principle of the strategy for Hg^{2+} detection is presented in Fig. 12. In brief, free Hg^{2+} specifically binding with the thymine–thymine

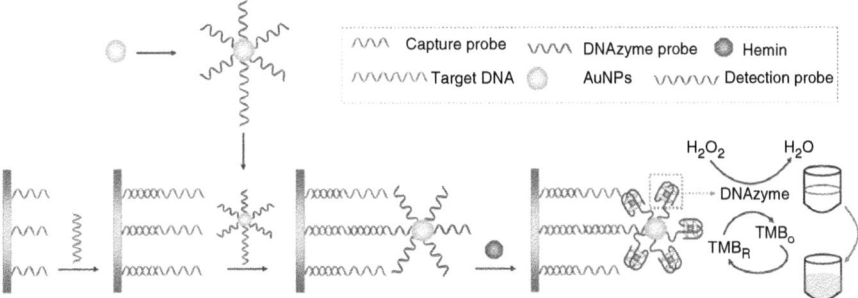

Fig. 11 Schematic representation of colorimetric sensing assay for *Salmonella* target DNA detection using DNAzyme probe self-assembled gold nanoparticles as a single tag (from Li et al. [41])

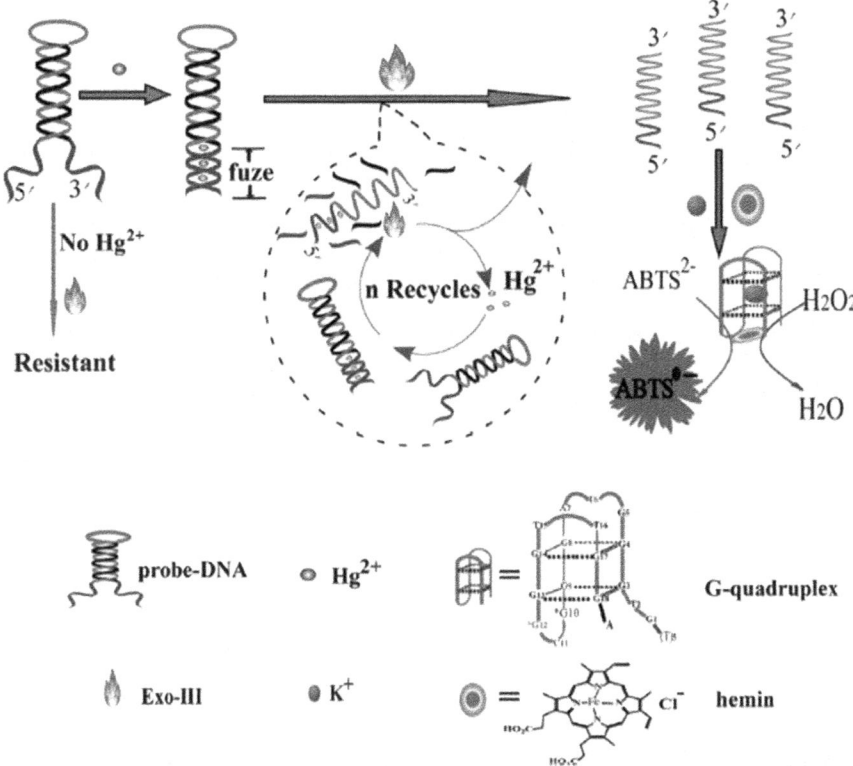

Fig. 12 Schematic illustration of the design principle of signal–colorimetric assay for Hg^{2+} detection based on Hg^{2+}-triggered Exo III-assisted target circulation and DNAzyme amplification. The "fuze" represents the duplex DNA with T-Hg^{2+}-T termini structure (from Ren et al. [46])

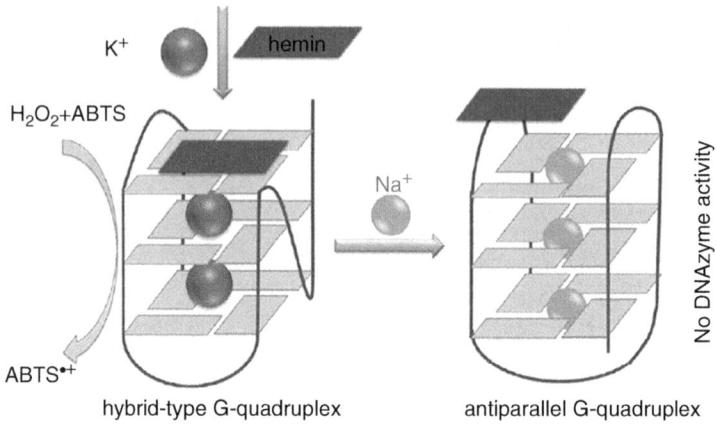

p25   5′-GGGCCAGGGAGCGGGGCGGAGGGGG-3′

Fig. 13 The schematic illustration for the Na⁺sensing mechanism (from Sun et al. [47]).

(T–T) mismatches termini of the hairpin DNA (H-DNA) could immediately trigger Exo III digestion of H-DNA duplex to set G-quadruplexes. Therefore, Hg^{2+}, which initiated new recycling, could produce cascades of DNAzyme in the presence of hemin. The corresponding DNAzymes efficiently catalyzed the H_2O_2-mediated oxidation of the $ABTS^{2-}$ to the colored product in the presence of hemin. This color change served as a signal output and could be instantly recognized by the naked eye and a chromometer.

Most G-quadruplexes are much more sensitive toward K^+ than toward Na^+; therefore, selective detection of Na^+ is difficult in the physiological system with a large amount of K^+, but there are specific G-quadruplexes exhibiting different conformations toward K^+ and Na^+. The G-quadruplexes with different conformations may have a different binding affinity toward hemin and thus show a difference in DNAzyme activity [47]. To design such a Na^+ sensor, Sun's group used the G-quadruplex (named p25), which shows a hybrid-type conformation with K^+, but an antiparallel conformation with Na^+ is developed to construct DNAzyme.

The DNAzyme of p25-hemin shows a great difference in catalytic activity with K^+ and Na^+ owing to the different binding affinity of hemin. Therefore, the catalytic activity of p25-hemin DNAzyme decreases with the G-quadruplex conformational transition as the colorimetric response of the $ABTS$-H_2O_2 system decreases (Fig. 13). Melamine (MA, 2,4,6-triamino-1,3,5-triazine, C3H6N6), an organic chemical inter-mediate heterocycle, can cause kidney toxicity and kidney stones in large amounts.

The European Community and other countries have established criteria for maximum residual limits of MA in a variety of foods, which are standard limits of 1 ppm (8 μM) for MA in infant formula and 2.5 ppm (20 μM) in other milk products [48]. Based on this demand, Wang et al. developed a colorimetric sensor of MA by exploiting the conformational change of hemin G-quadruplex-DNAzyme [49].

The principle of colorimetric detection is shown in Fig. 14. First, the hemin G-quadruplex-DNAzyme structure with multiple thymines is formed in the traditional conditions, which can catalyze H_2O_2-mediated TMB oxidation to colored products.

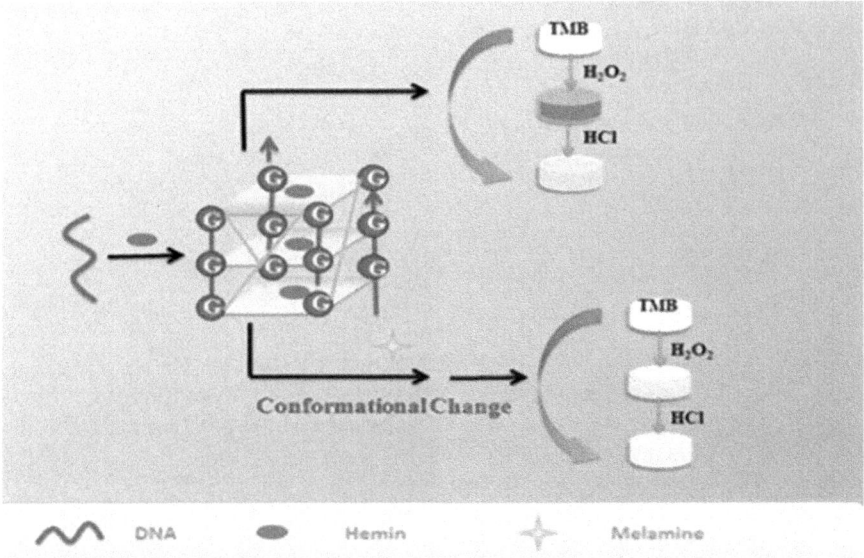

Fig. 14 The hemin G-quadruplex-DNAzyme for colorimetric and visual melamine detection (from Wang et al. [49])

When MA is added into the system, the conformation of the hemin G-quadruplex-DNAzyme is changed because of the formation of triple H-bonds between MA and the thymine in the structure of hemin G-quadruplex-DNAzyme. This, in turn, results in visual changes in TMB/H_2O_2 solution, which is the basis of colorimetric detection.

3.3   Chemiluminescence Sensors and Biosensors

Recently, CL has attracted much interest. Because of its inherent high sensitivity and wide linear working range with simple instrumentation and rapidity in signal detection, CL analysis, a very effective and well-accepted detection technology, has been successfully applied to trace quantities of analytes in various samples [50, 51]. Enzyme is the most popular strategy in CL immunoassay for signal amplification. Artificial enzymes, such as horseradish peroxidase (HRP)-mimicking hemin G-quadruplex DNAzymes, are attracting much attention from researchers in the CL detection of various types of analytes [52].

Typically, He's group designed a CL immunosensor for the detection of human leptin using a hemin G-quadruplex DNAzymes-assembled signal amplifier [53]. In this sensing platform, the primary antibody (anti-human leptin) was first immobilized on the 96-well plates, and human leptin and biotinylated secondary antibody were successively combined to form sandwich-type immune complex

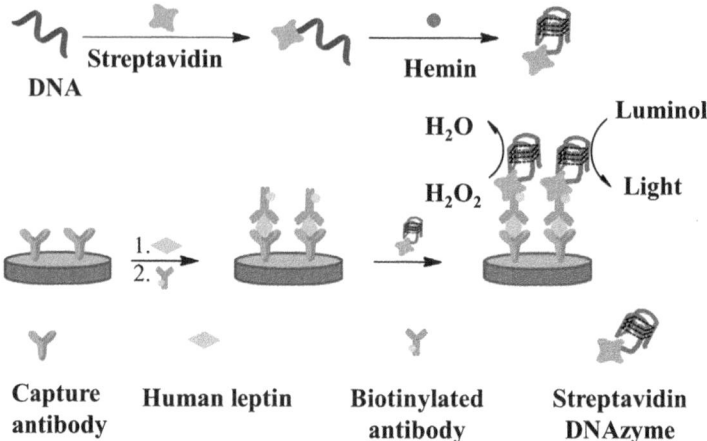

Fig. 15 Analytical procedure of DNAzymes-catalyzed chemiluminescent immunoassay (from He et al. [53])

(Fig. 15). Therefore, streptavidin labeled with hemin/G-quadruplex DNAzymes was assembled to the sandwich-type immunocomplex by streptavidin–biotin affinity. The DNAzymes showed a superior catalytic activity toward the CL reaction of luminol with hydrogen peroxide, resulting in a significant response signal.

Hun et al. further developed a CL assay for L-histidine based on controlled DNAzyme catalytic reactions on magnetic microparticles [54]. In this research, two oligomers, DNAzyme and substrate, were used (Fig. 16). The substrate was labeled with CL reagent. The DNAzyme was immobilized on the carboxylic magnetic beads and the substrate formed a stable duplex by allosteric, synergetic stabilization of each duplex by the other duplex. With the addition of L-histidine, the DNAzyme was activated and resulted in the formation of two fragments of labeled substrate that dissociate from the beads. After removal of the magnetic beads, the released fragment was detected with CL. This method was successfully utilized for the measurement of L-histidine concentration in real samples (human urine and serum). Thus, its application would be developed in fields such as biological research or diagnosis in clinical medicine.

Imaging array-based CL immunoassay of protein markers has attracted significant search efforts. The assay can be accomplished by the use of the CL-based immunosensor array and a charge-coupled device (CCD) camera. It has several advantages, such as high throughput, fast analysis, good selectivity, a wide linear range, and simple operation, without the need for external light source or optics [55, 56]. In this regard, Zong's group designed a CL immunosensor for cardiac troponin T by using silver nanoparticles functionalized with hemin G-quadruplex DNAzyme on a glass chip array [57]. The immunosensor array was fabricated by conjugating capture antibody (Ab$_1$) in 48 sensing cells on an aromatic aldehyde-functionalized disposable glass chip (Fig. 17). The tracing tag was prepared by assembling DNAzyme and detection antibody (Ab$_2$) on the surface of AgNPs. After

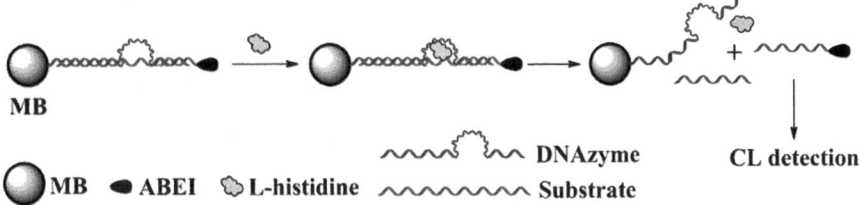

Fig. 16 Schematic diagram for the L-histidine biosensor fabrication based on L-histidine-controlled DNAzyme catalytic reactions coupled with magnetic separation (from Hun et al. [54])

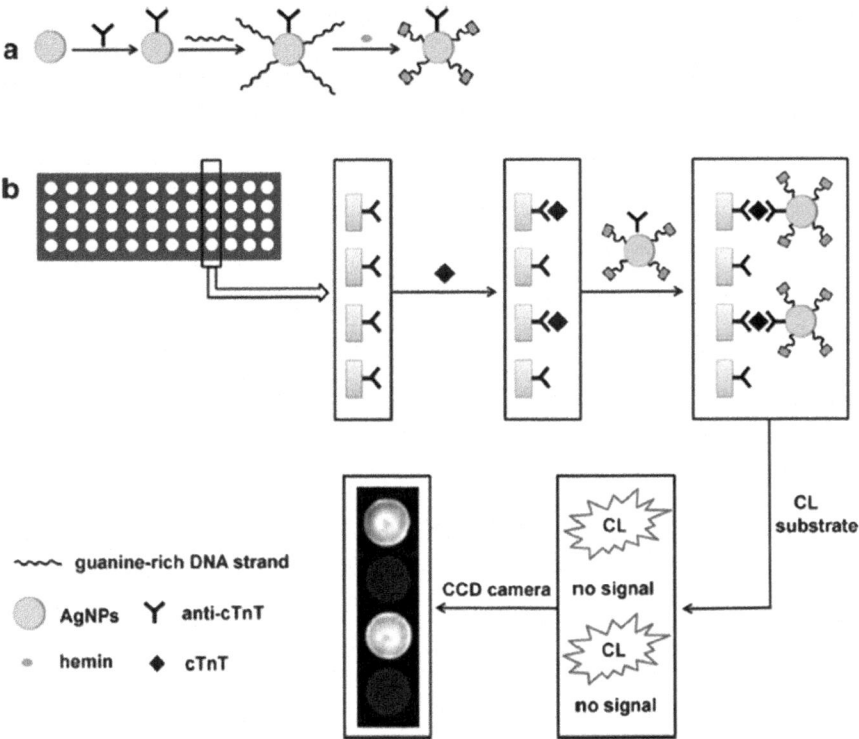

Fig. 17 Schematic diagram of (**a**) preparation of DNAzyme Ab2/AgNPs and (**b**) the immunosensor array with 4 × 12 cells and the process of highly sensitive chemiluminescent immunoassay of cTnT

formation of sandwich immunocomplex on the array, the tracing tag catalyzes the CL reaction of the luminol-p-iodophenol and H_2O_2 system to produce a CL signal, which is collected by a CCD camera.

It is known that G-quadruplexes are usually stabilized by coordination with alkaline cations, especially K^+ and Na^+ [58].

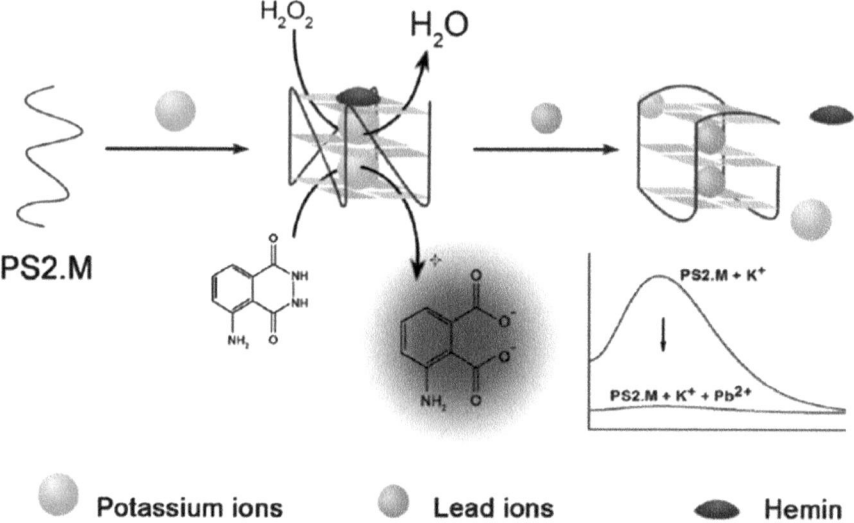

Fig. 18 Schematic illustration of Pb^{2+}-induced allosteric G-quadruplex DNAzyme

Considering that Pb^{2+} can induce a G-rich oligonucleotide (PS2.M) to form more compact quadruplex structures than K$^+$, and that the efficiency of Pb^{2+} in stabilizing the G-quadruplex is much higher than that of K$^+$, Wang and co-workers exploited these structural features to develop a detection method for Pb^{2+} based on its displacement of potassium in G-quadruplex DNAzyme [59].

As shown in Fig. 18, the K$^+$-stabilized G-quadruplex DNAzyme (with hemin as a cofactor) can effectively catalyze the oxidation of luminol by H$_2$O$_2$, giving rise to a CL signal. With the addition of Pb^{2+}, the DNAzyme deactivated and the CL intensity decreased through the substitution of K$^+$ in the DNAzyme by Pb^{2+}. The proposed sensor provides a highly sensitive strategy for Pb2+ detection, which has been successfully applied to lead detection in hair.

4 Conclusion and Perspective

In this chapter, we have summarized recent advances in DNAzyme-based sensing, focusing on electrochemical, colorimetric, and CL sensors. Over recent years, hemin G-quadruplex-based sensors have been a remarkable source of inspiration for chemists in their research, in view of their advantages over natural enzymes, such as thermal stability, high catalytic efficiency, easy preparation and modification, and lower cost than proteins. DNAzymes with peroxidase-like activity have been beneficial in a various bioanalytical applications, but still little research into optimizing sensing systems and eliminating matrix effects has been accomplished. For example, most sensors have been examined on model analytes in buffer solution under well-established laboratory settings; however, for a real diagnosis and environmental monitoring, matrix effects should be tested. Another challenge regarding DNAzyme is that there is a very limited number of substrates using readout. To expand the design flexibility of DNAzyme-based sensors, new signal transduction strategies are needed, such as colorimetric sensing, fluorescence, and electrochemistry. Because of amplified detection, satisfactory sensitivity, and straightforward operation of reported sensing systems, rapid progress and advancement of DNAzyme-based sensors are expected in the future.

References

1. Spanhel L, Weller H, Henglein A (1987) J Am Chem Soc 109:6632–6635
2. Neo JL, Kamaladasan K, Uttamchandani M (2012) Curr Pharm Des 18:2048–2057
3. Kosman J, Juskowiak B (2011) Anal Chim Acta 707:7–17
4. Breaker RR, Joyce GF (1994) Chem Biol 1:223–229
5. Ellington AD, Szostak JW (1990) Nature 346:818–822
6. Xiang GM, Jiang DN, Luo FK, Liu F, Liu LL, Pu XY (2014) Sens Actuators BChem 195:515
7. Gong L, Zhao Z, Lv Y-F, Huan S-Y, Fu T, Zhang X-B, Shen G-L, Yu R-Q (2015) Chem Commun 51:979–995
8. Li Y, Sen D (1997) Biochemistry 36:5589–5599
9. Travascio P, Li Y, Sen D (1998) Chem Biol 5:505–517
10. Suzuki M, Kobayashi S, Kimura M, Hanabusa K, Shirari H (1997) Chem Commun 2:227–228
11. King D, Wu J, Wang N, Yang W, Shen H (2009) Talanta 80:459–465
12. Juskowiak B (2011) Anal Bioanal Chem 399(9):3157–3176
13. Cheng X, Liu X, Bing T, Cao Z, Shangguan D (2009) Biochemistry 48:7817–7823
14. Kong D, Yang W, Wu J, Li C, Shen H (2010) Analyst 135:321–326
15. Travascio P, Witting PK, Mauk AG, Sen D (2001) J Am Chem Soc 123:1337–1348
16. Li Y, Sen D (2001) Biochemistry 36:5589–5599
17. Nakayama S, Sintim HO (2009) J Am Chem Soc 131:10320–10333
18. Kong D, Cai L, Guo J, Wu J, Shen H (2009) Biopolymers 91:331–339
19. Kong DM, Xu J, Shen HX (2010) Anal Chem 82:6148–6153
20. Xing Y, Zhong M, Han X, Yang X (2014) Int J Electrochem Sci 9:2978
21. La M, Feng Y, Yang C, Chen C (2014) Int J Electrochem Sci 9:6985
22. Hou T, Wang XZ, Liu XL, Pan C, Li F (2014) Sens Actuators B Chem 202:588
23. Liu L, Gao Y, Liu H, Xia N (2015) Sens Actuators B Chem. 208:137
24. Xia N, Liu L, Wu R, Liu H, Li S-J, Hao Y (2014) J Electroanal Chem 731:78
25. Willner I, Shlyahovsky B, Zayats M, Willner B (2008) Chem Soc Rev 37:1077
26. Hou L, Gao Z, Xu M, Cao X, Wu X, Chen G, Tang D (2014a) Biosens Bioelectron 54:365–371
27. Hou L, Tang Y, Xu M, Gao Z, Tang D (2014b) Anal Chem 86:8352–8358
28. Li Y, Deng J, Fang L, Yu K, Huang H, Jiang L, Liang W, Zheng J (2015) Biosens Bioelectron 63:1–6
29. Alizadeh N, Hallaj R, Salimi A (2017) Biosens Bioelectron 94:184–192
30. Lan T, Furuya K, Lu Y (2010) Chem Commun 46:3896–3898
31. Xu H, Xu P, Gao S, Zhang S, Zhao X, Fan C, Zuo X (2013) Biosens Bioelectron 47:520–523
32. Tang S, Tong P, Li H, Tang J, Zhang L (2013) Biosens Bioelectron 42:608–611
33. Liu S, Wei W, Sun X, Wang L (2016) Biosens Bioelectron 83:33–38
34. Golub E, Freeman R, Willner I (2011) Angew Chem Int Ed 123:11914–11918
35. Yuana Y, Gao M, Liu G, Chai Y, Wei S, Yuan R (2014) Anal Chim Acta 811:23–28
36. Stryer L, Berg JM, Tymoczko JL (2002) Biochemistry. W.H. Freeman, New York

37. Wang Z-H, Lu C-Y, Liu J, Xu J-J, Chen H-Y (2014) Chem Commun 50:1178–1180
38. Meng X, Zhou Y, Liang Q, Qu X, Yang Q, Yin H, Ai S (2013) Analyst 138:3409–3415
39. Li R, Zhen M, Guan M, Chen D, Zhang G, Ge J, Gong P, Wang C, Shu C (2013) Biosens Bioelectron 47:502
40. Wei W, Zhang DM, Yin LH, Pu YP, Liu SQ (2013) Spectrochim Acta A Mol Biomol Spectrosc 106:163–169
41. Li R, Jiang Q, Cheng H, Zhang G, Zhen M, Chen D, Ge J, Mao L, Wangab C, Shu C (2014) Analyst 139:1993–1999
42. Luo R, Li Y, Lin X, Dong F, Zhang W, Yan L, Cheng W, Ju H, Ding S (2014) Sensors Actuators B 198:87–93
43. Huang Y, Chen J, Zhao S, Shi M, Chen ZF, Liang H (2013) Anal Chem 85:4423–4430
44. Deng M, Zhang D, Zhou Y, Zhou X (2008) J Am Chem Soc 130:13095–13102
45. Zhang Q, Cai Y, Li H, Kong DM, Shen HX (2012) Biosens Bioelectron 38:331–336
46. Ren W, Zhang Y, Huang WT, Li NB, Luo HQ (2015) Biosens Bioelectron 68:266–271
47. Sun HX, Xiang J, Gai W, Liu Y, Guan AJ, Yang QF, Li Q, Shang Q, Su H, Tang YL, Xu GZ (2013) Chem Commun 49:4510–4512
48. Sun F, Ma W, Xu L, Zhu Y, Liu L, Peng C, Wang L, Kuang H, Xu C (2010) Trends Anal Chem 29:1239–1249
49. Wang G, Zhu Y, He X, Chen L, Wang L, Zhang X (2014) Microchim Acta 181:411–418
50. Li YP (2012) Microchim Acta 177:443–447
51. Han E, Ding L, Qian R, Bao L, Ju H (2012) Anal Chem 84:1452–1458
52. He Y, Sun J, Wang X, Wang L (2015) Sensors Actuators B Chem 221(31):792–798
53. He Y, Wang X, Zhang Y, Gao F, Li Y, Chen H, Wang L (2013) Talanta 116:816–821
54. Hun X, Xu Y, Bai L (2015) Microchim Acta 182:565–570
55. Liu AR, Zhao F, Zhao YW, Shangguan L, Liu SQ (2016) Biosens Bioelectron 81:97–102. https://doi.org/10.1016/j.bios.2016.02.049
56. Wang WW, Su XX, Ouyang H, Wang L, Fu ZF (2016) Anal Chim Acta 917:79–84. https://doi.org/10.1016/j.aca. 2016.03.001
57. Zong C, Zhang D, Yang H, Wang S, Chu M, Li P (2017) Microchim Acta 162:1–8
58. Wang H, Wang DM, Gao MX, Wang J, Huang CZ (2014) Anal Methods 6:7415–7419
59. Wang H, Wang DM, Huang CZ (2015) Analyst 140:5742–5747

Adv Biochem Eng Biotechnol (2020) 170: 107–120
DOI: 10.1007/10_2019_92
© Springer Nature Switzerland AG 2019
Published online: 8 March 2019

Aptazymes: Expanding the Specificity of Natural Catalytic Nucleic Acids by Application of In Vitro Selected Oligonucleotides

Johanna-Gabriela Walter and Frank Stahl

Contents

Abstract Aptazymes are synthetic molecules composed of an aptamer domain and a catalytic active nucleic acid unit, which may be a ribozyme or a DNAzyme. In these constructs the aptamer domain serves as a molecular switch that can regulate the catalytic activity of the ribozyme or DNAzyme subunit. This regulation is triggered by binding of the aptamers target molecule, which causes significant structural changes in the aptamer and thus in the entire aptazyme. Therefore, aptazymes function similar to allosteric enzymes, whose catalytic activity is regulated by binding of ligands (effectors) to allosteric sites due to alteration of the three-dimensional

J.-G. Walter (✉) and F. Stahl
Institute for Technical Chemistry, Leibniz University Hanover, Hannover, Germany
e-mail: walter@iftc.uni-hannover.de

structure of the active site of the enzyme. In case of aptazymes, the allosteric site is composed of an aptamer. Aptazymes can be designed for different applications and have already been used in analytical assays as well as for the regulation of gene expression.

Keywords Aptamer, Aptazyme, DNAzyme, Ribozyme

Abbreviations

ABTS	2,2′-Azino bis(3-ethylbenzthiazoline)-6 sulfonic acid
ALONA	Aptazyme-linked oligonucleotide assay
ATP	Adenosine triphosphate
cOligo	Complementary oligonucleotide
DNAzyme	Deoxyribozymes
FRET	Fluorescence resonance energy transfer
GMP	Guanosine monophosphate
HDV	Hepatitis delta virus ribozyme
HHR	Hammerhead ribozyme
IgE	Immunoglobulin E
LOD	Limit of detection
SELEX	Systematic evolution of ligands by exponential enrichment
VEGF	Vascular endothelial growth factor

1 Introduction

The term aptazyme was introduced by Robertson and Ellington in 1999 to describe the combination of an *apta*mer with a ribo*zyme* [1]. Aptazymes have been developed by rational design as well as by in vitro selection processes. While first aptazymes were composed of RNA aptamers and ribozymes, later the combination of DNA aptamers and DNAzymes paved the way to aptazymes with novel reactivities. Within the following sections, the structure elements of aptazymes, namely, the aptamer and the ribozyme/DNAzyme unit, will be briefly introduced; more in-depth information on ribozymes and DNAzymes can be found in other chapters of this volume. Subsequently, the mechanisms by which aptazyme activity is regulated are highlighted, and the diverse strategies to generate aptazymes will be pointed out. Finally various applications of aptazymes ranging from analytical applications to gene regulation will be summarized, and most recent achievements in these fields will be discussed as well as current limitations and future perspectives.

2 Components of Aptazymes

Aptazymes consist of an aptamer domain and a ribozyme or DNAzyme unit. Since these building blocks are the basis for the main characteristics of aptazymes, namely, their activity originating from the ribozyme or DNAzyme and the switchability of the activity which is controlled by the aptamer, a short introduction to these functional nucleic acids is given in the following.

2.1 Aptamers

Aptamers are either composed of RNA or DNA and can be generated in vitro via systematic evolution of ligands by exponential enrichment (SELEX) against a broad range of target molecules from small molecules to proteins and even cells. These oligonucleotides are able to adopt a defined three-dimensional structure that facilitates the binding of the target molecule via molecular recognition which is based on the complementarity of the surfaces of the aptamer and the target. Upon binding to their target, the predominant majority of aptamers undergoes considerable structural rearrangements known as adaptive binding or induced fit. This structure-switching property of aptamers is exploited in aptazymes: the role of the aptamer domain in aptazymes is to translate the presence of the aptamers target into conformational changes of the aptamer that affect the enzymatic unit, thereby altering its activity. Mainly two different modes of aptamer utilization can be distinguished in aptazymes: (1) The target-induced structural changes in the aptamer can directly be employed. In case of analytical applications, the mechanism is known as "target-induced structure switch" (Fig. 1a). For some aptazymes this mechanism is directly exploited to trigger the activity. (2) In other cases the fact that aptamers are oligonucleotides and can not only bind to their target but also to other oligonucleotides is used. Here it is possible to design oligonucleotides complementary to the section of the aptamer that is directly involved in binding of the target. Therefore, complementary oligonucleotide and target competitively bind to the aptamer, and binding of the target induces dissociation of the oligonucleotide and therefore a dramatic change of the conformation from a helical structure to a more sophisticated folded aptamer-target complex. In biosensor schemes this mechanism has been termed "target-induced dissociation of complementary oligonucleotides" (TID, Fig. 1b) and has shown to be especially useful for the detection of small molecules [2]. Within aptazymes it is also mostly used in combination with aptamers directed against small molecules and allows a more straightforward design of aptazymes when compared to the TISS mode.

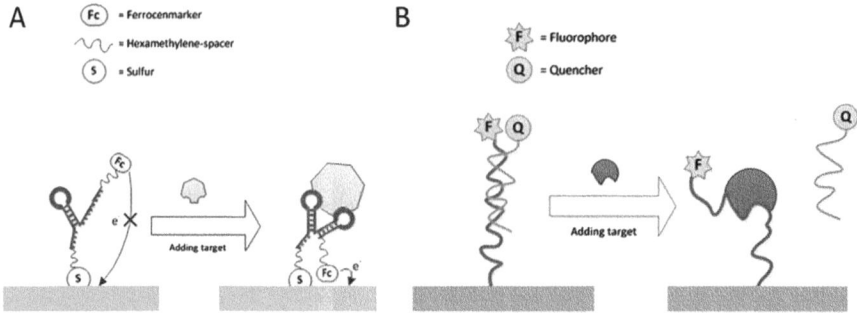

Fig. 1 Mechanisms by which target binding induces structural rearrangements of aptamers, exemplarily shown for aptasensor applications. (**a**) In target-induced structure switching (TISS), the adaptive binding induces conformational changes. (**b**) In target-induced dissociation (TID) of complementary oligonucleotides, the aptamer is hybridized with a complementary oligonucleotide in the off-state. Since the complementary oligonucleotide is designed to bind to the aptamers target-binding site, it can be released upon binding of the target. Modified from [2]

2.2 Enzymatic Units Used in Aptazymes

Aptazymes contain a catalytic active nucleic acid that can be either a ribozyme or a DNAzyme, two species that are in detail discussed in other chapters of this volume. Nonetheless, to allow the reader a facile understanding of aptazymes, the enzymatic nucleic acids will be briefly explained in this section.

In 1982 RNA molecules were discovered that are able to catalyze specific biochemical reactions. Due to their enzyme-like characteristics, they have been termed ribozymes. Most ribozymes catalyze cleavage or ligation of RNA or DNA, some also the formation of peptide bonds. The later reaction is related to the ribozymes role in the ribosome, where they participate in the linking of amino acids during protein biosynthesis. Ribozymes also are involved in various RNA processing reactions, such as RNA splicing, viral replication, and transfer RNA biosynthesis. This can be seen as the natural background of their ligation and splicing abilities.

The development of aptazymes, which combine ribozymes with RNA aptamers, was started in 1997 [3]. Various ribozymes have already been used in the development of aptazymes. These include hammerhead ribozyme, HDV, hairpin, group I intron, and X-motif ribozymes [4]. In Fig. 2 exemplarily, the hammerhead ribozyme (HHR) and the hairpin ribozyme are shown. While the natural HHR has no catalytic molecule, since it is consumed during the reaction by self-cleavage, it is possible to engineer variants that are composed of two strands, to obtain a catalytic RNA molecule. One strand contains the catalytic core and the substrate binding arms; the other represents the substrate of the HHR. During the reaction, the substrate strand interacts with the substrate binding arms via Watson-Crick base pairing and is cleaved. The fragments of the substrate RNA then dissociate from the complex, allowing for binding of new substrate and for multiple turnover.

Fig. 2 The hammerhead ribozyme (HHR). Reprinted with permission from [5]

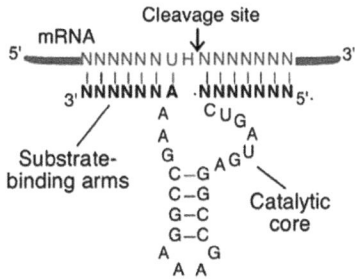

Hammerhead ribozyme

It is well known that some naturally occurring ribozymes are controlled in their activity by binding of specific molecules, which either act as a cofactor or as an allosteric modulator [6]. In this context, Lee et al. have discovered a group I self-splicing intron in *Clostridium difficile* that depends on cyclic di-GMP and regulates gene expression [7]. Although ligand-depending, and especially allosteric natural ribozymes, are rather rare today, it could be speculated that more will be revealed in the future [6].

One major disadvantage of ribozymes is their short half-life especially in the body. Here modifications of ribose can be used to stabilize the RNA molecule. In contrast to ribozymes, no naturally occurring deoxyribozymes (DNAzymes) have been observed up to now. DNAzymes have been identified which behave as ribonucleases or RNA ligases as well as catalyzing some other reactions. Besides the higher stability of DNA, other advantages of DNAzymes over ribozymes include the easier and more cost-effective production.

3   Mechanisms Involved in Aptazyme Function

Binding of aptamers and their targets in most cases is based on adaptive binding and includes massive structural rearrangement of the aptamers. In particular, loops and bulges that exhibit structural flexibility in the free aptamer acquire a defined conformation in the aptamer-target complex. This does not only affect the aptamer-binding site but also more adjacent domains, which can be DNAzymes or ribozymes in the case of aptazymes.

Binding of the aptamers target can either cause increase or decrease of aptazyme activities. When target binding causes activation of the aptazyme, the activity enhancement is typically $10-10^3$-fold [4]. Decrease of aptazyme activity upon target binding is rather rare; nonetheless there are some examples, as the ATP-regulated ribozyme published by Tang and Breaker [8]. They investigated an aptazyme composed of an ATP-binding aptamer and the hammerhead ribozyme (HHR) and found that binding of ATP results in a conformation in which steric interference occurs between aptamer domain and ribozyme unit. This results in 189-fold decrease of ribozyme activity [8].

When the aptamer is fused to a helical section of a ribozyme, target binding to the aptamer results in stabilization of this helix, thereby switching of the activity of the ribozyme occurs [9]. In many cases binding of the target results in structural stabilization of the active conformation of the ribozyme [10]. Aptamer targets that have already been used to allosterically regulate ribozyme activity include ATP, theophylline, flavin mononucleotide (FMN), cyclic nucleotide monophosphates, doxycycline, and pefloxacin [4]. DNAzymes have, e.g., been regulated by ATP [4].

4 Design of Aptazymes

Aptazymes can be designed either rationally, e.g., by combining an aptamer with desired specificity and a DNAzyme with a certain reactivity or by using combinational approaches as well as by combining both strategies. The design principles will be outlined in the following sections.

4.1 *Modular Rational Design of Aptazymes*

A straightforward method to design DNA aptazymes is to fuse the aptamer domain with a DNAzyme and to elongate the resulting chimera with an additional sequence complementary to the aptamers target-binding site (Fig. 3). This approach can be rationally designed and is easy to transfer to other aptazymes. It was introduced by Zhou et al. by combining two different aptamers (directed against ATP and interferon gamma) and two DNAzymes (8–17 DNAzyme and GR5 DNAzyme) [11]. The use of aptamers directed against interferon gamma is one of the rather rare examples where aptamers directed against proteins have been used to design aptazymes, while most of aptazymes still rely on aptamers directed against small molecules with a clear focus on ATP. The authors have directly fused the 5′terminus of the aptamer with the 3′terminus of the DNAzyme with no additional "helper sequences" required. In addition a short oligonucleotide was fused to the 5′terminus of the DNAzyme. This oligonucleotide was designed to be complementary to the target-binding region of the aptamer. This mechanism corresponds to the TID mechanism shown in Fig. 1b. In the absence of the target, the complementary oligonucleotide is hybridized with the aptamer, resulting in formation of a DNA hairpin structure, and formation of the active conformation of the DNAzyme is inhibited. In the presence of the target, the complementary oligonucleotide is dissociated from the aptamer resulting in more structural flexibility of the DNA chimera allowing the DNAzyme

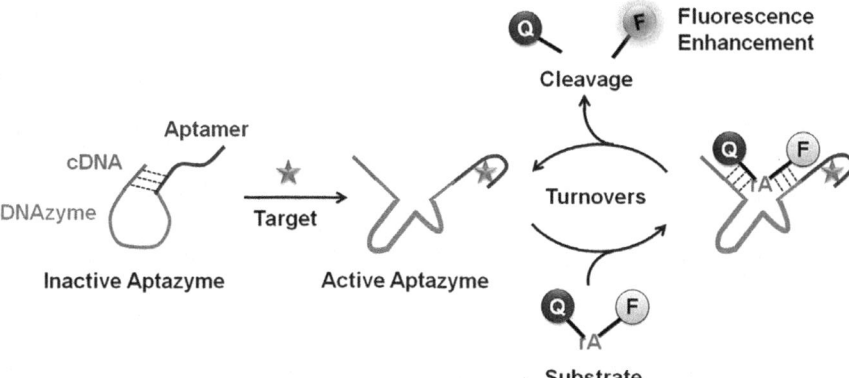

Fig. 3 Modular design of aptazymes by fusion of aptamers with DNAzymes and a sequence complementary to the aptamer. Reprinted from [11] with permission from Elsevier

to fold into its active conformation [6]. The active DNAzymes cleaved a substrate dual labeled with a fluorophore and a quencher, resulting in an increase of fluorescence in the presence of the target molecule. The aptazymes exhibit a limit of detection (LOD) for ATP in the μM range; interferon gamma was detectable with a LOD of 4.6 nM. Stability and specificity of the aptazymes were demonstrated by performing measurements in serum samples [11].

The main advantage of this combinational approach is that the DNAzyme and aptamer sequences do not need to be altered and the approach can be transferred to other combinations of aptamers and/or DNAzymes, thereby holding the potential for general applicability.

4.2 Combinational Approaches in Aptazyme Design

Aptazymes based on ribozymes are conventionally composed of three parts: the aptamer and the ribozyme and the so-called communication module that bridges the two aforementioned sections of the RNA aptazyme (Fig. 4) [3, 6]. In this setup, the communication module does not simply connect the two functional nucleic acids, but simultaneously exhibits a critical function. It is responsible for transmitting the ligand-binding from the aptamer domain to the ribozyme unit. Moreover careful design of the communication module is necessary, since it can determine whether binding of the ligand stabilizes or destabilizes the active conformation of the ribozyme [6].

Initially the setup of aptazymes as a communicator module-bridged fusion of aptamers and ribozymes was believed to be a modular system, in which each

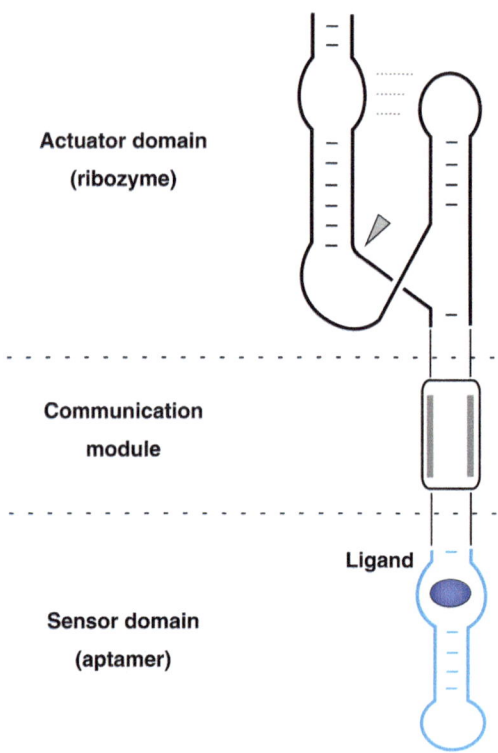

Fig. 4 Components of RNA aptazymes. The ribozyme (black) and the aptamer domain (blue) are connected with each other with a communication module (gray). Reprinted with permission from [6]

component (communicator module, aptamer, and ribozyme) is interchangeable. Later it was found that variation of the aptamer domain can in some cases result in failed activation of the ribozyme unit [1]. In these cases the combination of aptamer and ribozyme does not allow for efficient activation via the used communicator module. The need to find suitable communicator modules for different combinations of aptamers and ribozymes led to the development of in vitro methods for selection of communicator modules that are able to translate the binding event taking place at the aptamer into changes in ribozyme units' activity [4].

Combinational approaches have been developed in which a library of communication modules with randomized nucleotide regions is used. Here it is possible to select either communicator modules that inhibit the ribozyme in the presence of the aptamers ligand or modules that activate the ribozyme by careful design of selection and negative selection rounds [6]. Nonetheless, screening of the library is still laborious and time-consuming.

Modular rational design and combinations of modular design and combinational approaches are efficient methods to develop new aptazymes as discussed in detail by Felletti and Hartig [6].

5   Application of Aptazymes

5.1   Analytical Applications of Aptazymes

Aptazymes can be considered as ideal bioreceptor units for biosensor development, since they do not only provide specificity based on the aptamer properties but also facilitate signal generation and amplification based on the enzymatic unit.

Concerning the ribozyme- or DNAzyme-based signal generation, either ligand-dependent cleavage or ligation reactions can be exploited for readout. Moreover, fluorescence resonance energy transfer (FRET) or a combination of fluorophore and quencher can be used to generate analyte-dependent fluorescence signals. Since the catalytic nucleic acid unit of the aptazyme can turn round more than one substrate, they can function as signal amplification units.

For example, Pollet et al. have developed an aptazyme-linked oligonucleotide assay (ALONA), by combining a G-quadruplex aptazyme with an aptamer directed against immunoglobulin E (IgE) [12]. In the ALONA, anti-IgG antibodies were immobilized to capture IgE. The aptazyme binds to the IgE via the anti-IgE aptamer and is used as a reporter molecule. Bound aptazyme is amplified via Helicase-dependent amplification. Finally the amplified G-quadruplex structure incorporates hemin, which results in a HRP-like peroxidase activity [13]. This is used to oxidize 2,2′-azino bis(3-ethylbenzthiazoline)-6 sulfonic acid (ABTS) resulting in a color change of the solution. Thereby this approach combines even two modes of signal amplification, namely, the isothermal amplification of the aptazyme and the aptazymes potential to convert various substrate molecules. This sophisticated signal enhancement strategy is reflected by the achieved LOD of 1 pM IgE [12].

In a colorimetric approach, Wu et al. have developed a gold nanoparticle-based agglomeration assay [14]. They have combined an aptamer directed against VEGF with a DNAzyme. Gold nanoparticles were modified with 12 nt oligonucleotides via thiol coupling on the nanoparticle surface. A linker DNA was designed which did not only exhibit complementarity toward the oligonucleotides immobilized to the gold nanoparticles but also contains the cleavage site of the DNAzyme. In the assay the sample was incubated for 30 min with the aptazyme and the linker DNA, before the oligonucleotide-modified gold nanoparticles were added. In the presence of VEGF, the DNAzyme unit of the aptazyme was activated, resulting in the cleavage of the linker DNA. Consequently, the linker DNA is not able to cross-link the gold nanoparticles, and the solution remains red. In contrast, in the absence of VEGF, the linker DNA stays intact and cross-links the gold nanoparticles resulting in blue color based on nanoparticle agglomeration. Thereby, this assay allows for an easy readout, which can be either done by the naked eye or with a simple photometer. The LOD is 100 pM VEGF with a broad dynamic range of three orders of magnitude, and the authors suggest that the assay can be easily transferred to other protein targets, as long as suitable aptamers are available [14].

When thinking of aptazymes as molecular biosensors, in which the aptamer represents the bioreceptor and the enzymatic unit facilitates signal generation, the communication module or other portions of the aptazyme that translate the binding of the ligand to the aptamer domain into changes of ribozyme (or DNAzyme) activity can be seen as the transducer of this biosensor. This enables the application of aptazymes even without immobilization on a sensor substrate in homogeneous assays. Due to the introduction of synthetic ribozymes, signal generation does no longer rely on the reactions performed by natural catalytic nucleic acids [6]: For example, the theophylline-responsive Diels-Alderase aptazyme facilitates turn-round of a fluorescent anthracene, thereby decreasing fluorescence intensity in a theophylline-dependent manner [15].

The unique possibility to use aptazymes not only for specific binding of the aptamers target but simultaneously for signal generation and amplification based on the DNAzyme enables efficient and easy operation [14]. Moreover, when compared with protein-based enzymes, DNAzymes are more stable, easier to manipulate, and more economic to produce.

5.2   Sensing in Living Cells

Sensing within the cellular environment involves some complications such as that the sensing module has to enter the cells and has to remain intact and functional within the cellular environment. To achieve this goal, Yang et al. have developed an aptazyme that is composed of an ATP-binding aptamer and the 10–23 DNAzyme [16]. The aptazyme was immobilized on gold nanoparticles together with the DNAzyme substrate oligonucleotide [16]. The substrate oligonucleotide was labeled with a fluorophore, and the aptazyme strand, composed of ATP-binding aptamer and DNAzyme, was modified with a quencher. In the absence of ATP, aptazymes were in an inactive state, and fluorescence of fluorophores was quenched by the quencher molecule and the gold nanoparticle. In the presence of ATP, the aptamer domain of the aptazymes undergoes structural rearrangements that induce refolding of the DNAzyme to its active conformation. The active DNAzyme cleaves the substrate oligonucleotide, thereby releasing a fluorophore-modified oligonucleotide fragment resulting in increase of fluorescence intensity. Because the gold nanoparticles were modified with an excess of substrate oligonucleotides (in comparison to aptazymes), the ATP-activated aptazyme can cleave more than one substrate oligonucleotide, resulting in an increase of fluorescence signal [16]. The detection limit for ATP was found to be 200 nM. Cellular uptake of the modified gold nanoparticles was investigated for HeLa and SMMC-7721 cells, and a high internalization efficiency was observed (approx. 6.4×10^4 particles per cell) with most of the particles present in the cytosol. This method holds the potential to be transferrable to other aptazymes, thereby providing a platform for intracellular detection of various biomolecules.

5.3 Aptazymes in Regulation of Gene Expression

Ribozyme-based aptazymes can be useful tools for the regulation of gene expression. In 2002 Thomson et al. were the first to rationally design an artificial riboswitch based on a ribozyme [17]. Here an aptamer directed against theophylline was introduced into group I intron resulting in theophylline-dependent splicing. Also in vitro selection approaches have been used to develop aptazymes for gene expression, as the group of Suess has shown for a tetracycline-dependent aptazyme [18].

Besides showing the desired activity in vitro, gene expression regulation requires stability and activity of aptazymes in vivo, a goal which is hard to achieve [17]. Difficulties may arise either from intracellular conditions, which hardly can be reproduced in vitro, RNA folding paths that may differ in vivo and in vitro, and the RNA stability in the cellular environment. To address the influence of intracellular environment, in vivo screening approaches have been developed, in which the activity of the ribozyme-based aptazyme is coupled to a reporter gene. For instance, the group of Harting has created a library composed of HHR and the theophylline-binding aptamer connected by a randomized communication module [19]. eGFP was used as a reporter gene to easily access aptazyme activity after transformation into *E. coli.*

The use of ligand-dependent aptazymes in gene regulation holds potential in biotechnological applications as well as in therapeutic strategies [6]. The main advantage is that gene expression can be controlled externally by addition of the corresponding ligand. The advantage of ribozyme-based aptazymes in gene expression is their modular nature that – at least in principle – allows for easy adaption to different applications by combining respective aptamers and ribozymes. Moreover – when thinking of the future therapeutical use of aptazymes – the RNA basis should be beneficial over protein-dependent mechanisms of gene regulation due to RNA's lack of immunogenicity. Since the spectrum of available aptamer domains that are functional in vivo and respond to a suitable biocompatible ligand is still limited today, future research will focus on the development of novel aptamers for this application and the integration of these aptamers into aptazymes [6].

More recently, aptazymes have been also exploited for genome editing. Tang et al. have combined self-cleaving RNA with guide RNA, thereby designing aptazyme-embedded guide RNAs that enable ligand-controlled editing in a temporal and spatial controlled manner [20].

6 Current Limitations and Future Perspectives

Obviously there are two major types of limitations in the context of aptazyme development: The first relates to the limited availability of the aptazymes building blocks, namely, aptamers and RNA or DNAzymes. While aptamers are straightforward to select via SELEX to obtain aptamers with desired specificity, the amount of available RNA or DNAzymes is still restricted, resulting in a limitation of the types of reactions that can be catalyzed by aptazymes. The second source of limitations is the combination of aptamer and enzymatic unit to a functional aptazyme, which often involves cumbersome trial and error approaches or elaborate screening processes.

Based on the restrictions concerning the reactions that can be catalyzed by aptazymes, the focus of aptazyme application will most likely be within analytical applications, where the same catalyzed reaction can be used for the detection of various analytes by incorporation of different aptamers. Here, the catalyzed reaction can be exploited for signal amplification resulting in high sensitivity. Besides analytical applications, the range of reactions performed by RNA and DNAzymes can be exploited in manipulation of RNA and DNA, as outlined for regulation of gene expression.

In the context of intracellular applications, one drawback of aptazymes might be their negative charge. This is known to impede with crossing of cellular membranes [9]. Nonetheless, due to the improvements of intracellular delivery of drugs, including therapeutic nucleic acids, and aptamer-based delivery systems, this issue will not be a final restriction.

7 Conclusions

Aptazymes have already been used in various applications. The nucleic acid structure of aptamers, ribozymes, and DNAzymes allows for their easy production and modification, which is especially useful when designing aptazymes for analytical applications, where fluorophores or other modifications can be introduced. Moreover, the nucleic acid structure also allows for the use of combinational methods to develop and/or optimize aptazymes. In analytical applications aptazymes can result in extraordinary sensitivity, since they can turn over multiple substrate molecules, thereby providing signal amplification.

References

1. Robertson MP, Ellington AD (1999) In vitro selection of an allosteric ribozyme that transduces analytes to amplicons. Nat Biotechnol 17(1):62–66
2. Walter J et al (2012) Aptasensors for small molecule detection. Z Naturforsch 67(b):976–986
3. Tang J, Breaker RR (1997) Rational design of allosteric ribozymes. Chem Biol 4(6):453–459
4. Silverman SK (2003) Rube Goldberg goes (ribo)nuclear? Molecular switches and sensors made from RNA. RNA 9(4):377–383
5. Sano M, Kato Y, Taira K (2005) Functional gene-discovery systems based on libraries of hammerhead and hairpin ribozymes and short hairpin RNAs. Mol BioSyst 1(1):27–35
6. Felletti M, Hartig JS (2017) Ligand-dependent ribozymes. Wiley Interdisci Rev RNA 8(2)

7. Lee ER et al (2010) An allosteric self-splicing ribozyme triggered by a bacterial second messenger. Science 329(5993):845–848
8. Tang J, Breaker RR (1998) Mechanism for allosteric inhibition of an ATP-sensitive ribozyme. Nucleic Acids Res 26(18):4214–4221
9. Famulok M, Hartig JS, Mayer G (2007) Functional aptamers and aptazymes in biotechnology, diagnostics, and therapy. Chem Rev 107(9):3715–3743
10. Soukup GA, Breaker RR (1999) Design of allosteric hammerhead ribozymes activated by ligand-induced structure stabilization. Struct Folding Des 7(7):783–791
11. Zhou ZJ et al (2015) A general approach for rational design of fluorescent DNA aptazyme sensors based on target-induced unfolding of DNA hairpins. Anal Chim Acta 889:179–186
12. Pollet J, Strych U, Willson RC (2012) A peroxidase-active aptazyme as an isothermally amplifiable label in an aptazyme-linked oligonucleotide assay for low-picomolar IgE detection. Analyst 137(24):5710–5712
13. Kong DM et al (2010) Structure-function study of peroxidase-like G-quadruplex-hemin complexes. Analyst 135(2):321–326
14. Wu D et al (2016) Colorimetric detection of proteins based on target-induced activation of aptazyme. Anal Chim Acta 942:68–73
15. Helm M et al (2005) Allosterically activated Diels-Alder catalysis by a ribozyme. J Am Chem Soc 127(30):10492–10493
16. Yang YJ et al (2016) Aptazyme-gold nanoparticle sensor for amplified molecular probing in living cells. Anal Chem 88(11):5981–5987
17. Thompson KM et al (2002) Group I aptazymes as genetic regulatory switches. BMC Biotechnol 2(21):12
18. Wittmann A, Suess B (2011) Selection of tetracycline inducible self-cleaving ribozymes as synthetic devices for gene regulation in yeast. Mol Biosyst 7(8):2419–2427
19. Wieland M, Hartig JS (2008) Improved aptazyme design and in vivo screening enable riboswitching in bacteria. Angew Chem Int Ed 47(14):2604–2607
20. Tang WX, Hu JH, Liu DR (2017) Aptazyme-embedded guide RNAs enable ligand-responsive genome editing and transcriptional activation. Nat Commun 8:15939

Index

Printed by Printforce, the Netherlands